INTERNATIONAL OCCUPATIONAL SAFETY AND HEALTH RESOURCE CATALOGUE

Jane H. Ives

PRAEGER SPECIAL STUDIES • PRAEGER SCIENTIFIC

Library of Congress Cataloging in Publication Data

Ives, Jane H.
  International occupational safety and health resource catalogue.

  Bibliography: p.
  Includes indexes.
  1. Industrial hygiene—Societies, etc.—Directories.
2. Industrial safety—Societies, etc.—Directories.
I. Title.
RC967.I86   363.1'1'025   81-13823
ISBN 0-03-060299-8   AACR2

Published in 1981 by Praeger Publishers
CBS Educational and Professional Publishing
A Division of CBS, Inc.
521 Fifth Avenue, New York, New York 10175 U.S.A.

© 1981 by Praeger Publishers

All rights reserved

123456789 145 987654321

Printed in the United States of America

ACKNOWLEDGMENTS

The International Conference on the Exportation of Hazardous Industries, Technologies and Products to Developing Countries and the International Occupational Safety and Health Resource Catalogue were created through the efforts of many people. I would like to thank Jeanne Reisman, Nicholas A. Ashford, Daniel Berman, Barry Castleman, Tish Davis, Denny Dobbin, Stanley W. Eller, Ray Elling, Robert Hayden, Brett Hill, Helen Jacobs Ives, Dieter Koch-Weser, Charles Levenstein, Joan Nicholson, Tony Robbins, Les Trachtman, Joe Valesquez, Frank Wallick, Jim Weeks, and Margaret Wood for their support, encouragement, and technical assistance.

PREFACE

This Catalogue was initiated at the International Conference on the Exportation of Hazardous Industries, Technologies, and Products to Developing Countries, which was held in New York City, November 2 and 3, 1979. The export of hazards from the United States to Third World countries to avoid the constraints imposed by occupational safety and health standards and environmental regulations were the major issues of discussion at the conference.

The Catalogue includes information from other sources on the listings of resources in the areas of environmental quality and occupational safety and health concerns. The Catalogue consists of four sections: a list of organizations, a bibliography, a series of appendixes providing additional information on the subject areas, and indexes to the organizations by both name and subject.

The list of organizations was compiled by means of a survey questionnaire that started as a handout at the conference but soon became the instrument of a worldwide mail and telephone survey. The listing for each organization includes basic identifying information, major goals and objectives, research efforts and programs, and publications. Organizations were selected on the basis of their expressed interest in environmental quality, hazard export, or occupational safety and health.

Considerable effort was expended to make the list comprehensive. A few agencies and organizations did not return the survey questionnaire, and others were doubtless missed in the process of networking on which the identification of worldwide resources depends. It is our intention to update and refine this Catalogue as more information becomes available. We urge all those who share our concerns to forward to us any additional information about groups, research, programs, or publications that should be included in an update.

We should also note that the compilation of this international Catalogue, based on a mail questionnaire, presents a number of editorial problems difficult to resolve. Many agencies did not fill out the questionnaire, but simply sent us publications or brochures in the hope that we could extract the appropriate information. Additionally, materials were presented informally so that the details necessary for accurate citation were missing.

We have presented our best reading of the material we received, striving for as much accuracy and consistency in format as possible.

The sheer quantity of response to this initial effort shows that concern for these issues is deeply and broadly shared around the world. We hope that this Catalogue will serve as a catalyst to promote efforts to research and develop strategies for protecting health, safety, and environmental quality in the world community.

CONTENTS

	Page
ACKNOWLEDGMENTS	v
PREFACE	vi
INTRODUCTION Nicholas A. Ashford	x

INTERNATIONAL OCCUPATIONAL SAFETY AND HEALTH
RESOURCES AND ORGANIZATIONS—BY GEOGRAPHIC AREA

ORGANIZATIONAL GUIDE	3
AFRICA	3
ASIA	6
AUSTRALIA	15
CENTRAL AMERICA	17
EUROPE	18
MIDDLE EAST	59
NORTH AMERICA	61
Canada	61
Mexico	74
United States	76
SOUTH AMERICA	164
SOUTH PACIFIC	170
USSR	179
BIBLIOGRAPHY	181

Page

APPENDIXES

A	DATA COLLECTION METHODOLOGY	189
B	SAMPLE LETTER AND QUESTIONNAIRE	190
C	INTERNATIONAL CONFERENCE ON THE EXPORTATION OF HAZARDOUS INDUSTRIES, TECHNOLOGIES, AND PRODUCTS TO DEVELOPING COUNTRIES	192
D	COMMITTEES ON OCCUPATIONAL SAFETY AND HEALTH (COSH GROUPS)	195
E	LABOR EDUCATION PROGRAMS IN THE UNITED STATES	202
F	EDUCATIONAL RESOURCES CENTERS IN THE UNITED STATES	204
G	OSHA NEW DIRECTIONS PROGRAMS	206
H	ORGANIZATIONAL LISTING OF NATIONAL AND INTERNATIONAL UNION CONTACTS, SELECTED AFL-CIO DEPARTMENTS, AND FEDERAL ADVISORY COUNCIL ON OCCUPATIONAL SAFETY AND HEALTH	224
I	A GUIDE TO WORKER EDUCATION MATERIALS IN OCCUPATIONAL SAFETY AND HEALTH	235
J	INTERNATIONAL ORGANIZATIONS	285
K	NATIONAL INSTITUTE FOR OCCUPATIONAL SAFETY AND HEALTH	295
NAME INDEX		299
SUBJECT INDEX		302
ABOUT THE AUTHOR		311

INTRODUCTION

Nicholas A. Ashford

During the past decade there has been an unprecedented increase in the concern for occupational and environmental health and the need for monitoring and controlling private-sector activities associated with industrial production and agriculture. This concern has taken on important multinational and international dimensions. The transfer of technology between developed and developing countries has resulted in both the increased export of hazardous products, such as pesticides, and the creation of new industrial production facilities in developing countries. Technology transfer that creates occupational and environmental problems has serious economic and political consequences for both donor and recipient countries.

While the U.S. regulatory scheme is the most highly developed domestic control system, little control of technology transfer is addressed by current law. Stringent U.S. regulations are alleged to create a competitive disadvantage to U.S. firms in international markets; at the same time, foreign countries claim that U.S. regulations are a trade barrier to their entering U.S. markets. There is a need for the development of uniform national policies and international agreement for the control of occupational and environmental hazards in order to prevent pollution havens and select victimization of workers, consumers, and citizens of countries with a need to undergo rapid improvement in their standard of living. The key to effectuating uniform protective practices on an international basis is the existing national manpower and institutional resources. This Catalogue represents a beginning at identifying those resources in both developed and developing countries.

International Occupational Safety and Health Resources and Organizations: by Geographic Area

ORGANIZATIONAL GUIDE

The list is organized by the following major geographical areas:
- Africa
- Asia
- Australia
- Central America
- Europe
- Middle East
- North America
- South America
- South Pacific
- USSR

The countries are organized alphabetically within each major geographic area. The United States and Canada are further subdivided by states and provinces, respectively, which are also listed alphabetically.

Each organization is designated as belonging to one of the following seven categories:
- Academic
- Government
- Health
- International
- Labor
- Private/Industrial
- Public Interest

Each organizational listing includes (where applicable):
- Organization Name
- Address
- Telephone/Telex
- Contact Person
- Objectives/Goals
- Research/Projects
- Publications Available

AFRICA

Kenya

International

Environmental Liaison Centre (ELC)
P. O. Box 72461
Nairobi, Kenya
Telephone: 24770

4 / SAFETY AND HEALTH CATALOGUE

Contact:
>Delmar Blasco, Executive Officer

Objectives/Goals:
>Established by nongovernmental organizations (NGOs) as a communication link with the United Nations Environment Program (UNEP) and the UN Centre for Human Settlements, both headquartered in Nairobi, on behalf of NGOs around the world; and to work with NGOs in developing countries. Its purposes are to protect the earth's ecosystems for human health and well-being; to promote the wise management and balanced distribution of resources; and to improve human settlements.

Research/Projects:
>Provides conference facilities and services for NGOs in Kenya.

>Conducts yearly NGO profile surveys concerning all matters related to the environment.

>ELC Referral Service.

Publications:
>ECO Forum, bimonthly newsletter. It functions as a "window" on UNEP and the Human Settlements Centre analyzing programs, describing past and future meetings, noting personnel changes, reporting on Government Council decisions, and generally advising NGOs on positions being taken by the two UN agencies.

>The NGO World Environment Day Information Pack in three languages (French, English, and Spanish).

>EE Switchboard, an NGO environmental education newsletter.

>The Quest for Harmony: Perspectives on the New International Development Strategy, Calestous Juma (Kenya: Environment Liaison Centre, August 1980).

Sudan

Government

Ministry of Health
Occupational Health Department
P. O. Box 303
Khartoum, Sudan

Contact:
   Dr. Yousif Osman, Director-General

Objectives/Goals:
   Environmental and health evaluation and control.

   Delivery of occupational health services to workplaces.

   Training in occupational safety and health concerns.

   Research.

Research/Projects:
   Guidelines for Education in Occupational Health for Migrant Workers.

   Studies on the Thermal Environment in Various Factories in Sudan.

   Environmental Studies in Textile Mills.

   Lead Problems in Electric Accumulation.

Publications:
   Statement on Air Pollution in Sudan.

   Exposure to Pesticides in Agriculture: A Survey of Spraymen Using Dimethoate in Sudan.

   Exposure to Pesticides of Workers Loading Planes for ULV Spraying in Gezira Scheme, Sudan.

   Occupational Health Problems in Developing Countries.

   Guidelines to Education in Occupational Health for Migrant Workers.

ASIA

Bangladesh

Public Interest

Gonoshasthaya Kendra
P. O. Box Nayarhat, via Dhamrai
Dacca, Bangladesh
Telex: SAVAR 80/Gramgoro, Dacca

Contact:
    Dr. Zafrullah Chowdhury, Project Coordinator

Objectives/Goals:
    To contribute to the liberation of the villagers through projects primarily in the areas of health, education, and economic independence.

Research/Projects:
    Preventive/curative health programs.

    Family planning with research in the area of Depoprovera.

    Education and experimental school for the children of landless and marginal farmers.

    Conscientization programs for credit union members and village women.

    Vocational training programs for men and women (shoe factories, bakeries, sewing/handicrafts, woodwork, and metal workshop).

Publications:
    Publications Department for development of teaching aids and health bulletins. Extensive list available upon request.

China

Public Interest

Chinese Academy of Medical Sciences
Institute of Health

Labor Hygiene Department
Nan Wei Road
29, Beijing, China

Objectives/Goals:
> To study the hazardous effects, health criteria, and standards of harmful factors in the workplace.
>
> To study the methodology for the determination of harmful substances in the workplace.
>
> To study the appropriate drugs for the prevention and treatment of occupational diseases and industrial poisoning.

Research/Projects:
> Diagnosis and treatment of silicosis; biochemical diagnosis (ceruloplasmin, lysozyme) and drugs (low mol. wt. P204, aluminum compounds, Chinese herbs) for treatment.
>
> Surveys on high-temperature working conditions and preventive measures.
>
> Prevention and treatment of allyl chloride poisoning; electromyogram for early diagnosis; drugs and Chinese herbs for treatment.
>
> Diagnosis, treatment, and prevention of benzene poisoning; biochemical diagnosis (LAP) and Chinese herbs for treatment.

Publications:
> <u>Chinese Magazine for Preventive Medicine and Special Reports</u>.

Hong Kong

Public Interest

Asia/North America Communications Center
110 Terrace Blvd., Room 304
New Hyde Park, New York 11040
Telephone: (516) 437-6064

Contact:
> Thomas P. Fenton

8 / SAFETY AND HEALTH CATALOGUE

Asia/Pacific
2 Man Wan Road
17-C
Kowloon, Hong Kong
Telephone: 3-035271

Contact:
    John Sayer; Christine Vertucci

Objectives/Goals:
    Hong Kong-based nonprofit Research and Documentation Center. Major objective is to systematically gather, organize, analyze, and publish information on the U.S. economic, political, and military involvement in Asia.

Research/Projects:
    Current Projects: Compilation of descriptive data on all transnational corporations active in the Asia region in collaboration with the United Nations Center on Transnational Corporations. Data include parent and affiliate names; affiliate addresses; line of business; sales; capitalization, assets; historical information; and industry breakdown on all Asian-based affiliates.

    Past Projects: Asia in America, prepared for the UNITAR meeting in New Delhi, India, March 11-18, 1980. A statistical summary of U.S. direct investments in Asia. It includes the most current and complete collection of published and unpublished data available from the U.S. Department of Commerce on U.S. direct investments abroad.

    Prospective Projects: A major study of health and safety-related issues in Asia. The focus will be on one industry (for example, chemical or asbestos industry) in one particular country as an illustration of conditions in U.S. multinational corporations throughout Asia, to begin Fall 1980.

Publications:
    Asia Monitor, quarterly digest of U.S./Asia-related economic news.

    America in Asia: Research Guides on United States Economic Activity in Pacific Asia: Vol. I, descriptive listing of sources of information in Asia on U.S. economic involvement in Asia.

America in Asia: A Handbook of Facts and Figures on United States Economic and Military Activity in Pacific Asia: Vol. II, sourcebook of data on U.S. economic and military power in Asia.

A Survey of Education/Action Resources on Multinational Corporations, an extensive bibliography of resources on multinational corporations.

Young Workers Confederation
Young Workers Centre
23 Block 28
Sall Mall Pink
Kowloon, Hong Kong
Telephone: 3-350812

Contact:
> Cheung Yuk-King, Chairperson

Objectives/Goals:
> To work toward the education and organization of young workers.
>
> To raise the consciousness of workers to the injustices in their work and living situations.
>
> To assist in the education and formation of labor leaders and worker groups.
>
> To encourage workers to protect their human rights.
>
> To share concern for workers with church people, social agencies, and other labor groups.

Research/Projects:
> Industrial Safety Exhibition and Booklet, 1979.

Publications:
> Young Workers Monthly.
>
> > Young Workers Center: Labour Education (slides and tapes)

10 / SAFETY AND HEALTH CATALOGUE

India

Academic

Centre for Education and Documentation
3 Suleman Chambers
4 Battery Street
Bombay, India
Telephone: 400 049 India

Contact:
    Mr. Pradeep Guha, General Secretary

Objectives/Goals:
    Documentation and social education.

Research/Projects:
    Planned study on multinational corporations in the drugs and pharmaceuticals industry in India.

Publications:
    List available upon request.

Public Interest

Indian Association for Water Pollution Control
c/o NEERI, Nehru Marg
Nagpur 440 020, India
Telephone: 26071; Telex: 233

Contact:
    V. Raman/P.V.R.C. Panicker, Secretary/Treasurer

Objectives/Goals:
    Advancement and promotion of knowledge and practical know-how in water pollution control.

Research/Projects:
    Information available upon request.

Publications:
    Convention Volume/Technical Annual.

Management Development Institute
F45, N.D.S.E.I.
New Delhi, India 110 049
Telephone: 31-4519

Contact:
    Amitav Rath

Objectives/Goals:
    Training and research on problems of management and economic development.

Research/Projects:
    Studies on the hotel industry, textiles, mining, small-scale industries, multinational companies, energy, technology, and so on.

Publications:
    List available upon request.

Japan

Government

National Institute of Industrial Health
21-1, Nagao 6-chome
Tama-ku, Kawasaki 312
Japan
Telephone: (044) 865-6111

Contact:
    Hiroyuki Sakabe, M.D., Director

Objectives/Goals:
    Subordinate organization in the Ministry of Labor, established to conduct interdisciplinary research on the maintenance and promotion of workers' health and the investigation, diagnosis, and prevention of occupational diseases.

Research/Projects:
    Research activities in industrial physiology, occupational diseases, experimental toxicology, industrial epidemiology, environmental hygiene, and human-environmental engineering.

Publications:
> Industrial Health, published quarterly in English.
>
> Annual report (in Japanese).

Public Interest

Pacific-Asia Resources Center
P. O. Box 5250
Tokyo International
Japan
Telephone: (03) 291-5901

Contact:
> Muto Ichiyo

Objectives/Goals:
> Research on Japanese transnational corporations and their activities in the Third World.

Research/Projects:
> Information available upon request.

Publications:
> AMPO (quarterly magazine).
>
> ASIA News (monthly), covering current situation in Japan and Asia.
>
> Slides on Japanese pollution "export" to the Third World.

Republic of Singapore

Academic

University of Singapore
Department of Social Medicine and Public Health
Outram Hill
Singapore 0316
Republic of Singapore
Telephone: 2226444/5; Telex: UNIVSPORE

Contact:
> Professor W. O. Phoon, Head of Department

Objectives/Goals:
>To conduct educational programs in occupational health and safety for university undergraduates, postgraduates, other health personnel, and the public.
>
>To conduct research in occupational health and safety.
>
>To offer a consultancy and advisory service to work sites on occupational health and safety.

Research/Projects:
>Decompression illness among deep-sea fishermen divers.
>
>The health status and occupational hazards of firemen.
>
>Metabolism of lead and its compounds.
>
>Occupational health in small factories.

Publications:
>Scientific papers are published regularly in local and international journals. Information available upon request.

Government

Industrial Health Division
Ministry of Labour
5 Halifax Road
Singapore 0922
Republic of Singapore
Telephone: 2538388

Contact:
>Dr. P. K. Chew, Director

Objectives/Goals:
>To investigate work sites and evaluate industrial hygiene conditions in Singapore.
>
>To research and evaluate methods for controlling and preventing occupational diseases.

Research/Projects:
>Mass hysteria in industries.

Relationship between noise and high blood pressure in industrial workers.

A survey of hazards.

Publications:
Handbook of Notifiable Industrial Diseases in Singapore.

A Guide to the Assessment of Traumatic Injuries for Workmen's Compensation.

Proceedings of the Regional Seminar on Occupational Health and Ergonomic Applications in Safety Control organized by the Society of Occupational Medicine, Singapore, September 18-21, 1978.

Thailand

Government

The Ministry of Public Health
Department of Health
Occupational Health Division
Devavesm Palace
Bangkok-2, Thailand
Telephone: 2825176

Contact:
Dr. Chinosoth Husbumrer

Objectives/Goals:
Prevention of occupational diseases through biological and environmental monitoring programs.

Research/Projects:
Occupational health problems in industry and agriculture.

Permissible duration of exposure to occupational hazards.

Publications:
Thailand Journal of Health and Environment.

AUSTRALIA

Academic

The University of Sydney, N.S.W.
School of Public Health and Tropical Medicine
Sydney, Australia 2006
Telephone: (02) 6604555

Contact:
    Dr. T. Ng, Senior Lecturer in Occupational Health

Objectives/Goals:
    Teaching, research, and consultation in environmental health and safety.

Research/Projects:
    Occupational factors in coronary heart disease and hypertension.

    Health and atomic energy research workers.

    Mesothelioma register.

Publications:
    1978-79: Survey of health of employees in the research establishment of the Australian Atomic Energy Commission, Sydney: First Report—Analysis of Medical Interview Data; Second Report—Interpretation of Hazard and Recommendations.

    1973: Control of Coronary Heart Disease in the Australian Post Office, Occupational Health Section, School of Public Health and Tropical Medicine.

Government

Health Commission of New South Wales
Division of Occupational Health
Lidcome Hospital
Lidcome, N.S.W. 2141, Australia
Telephone: (02) 646-0379

Contact:
    Dr. W. A. Crawford, Director

Objectives/Goals:
    The diminution of ill health and injuries induced by the nature and conditions of work.

Research/Projects: Studies on the following:

    The effects of lead and other heavy metals, dusts, solvents, asbestos, herbicides, pesticides, ionizing and nonionizing radiation.

    In-depth respiratory and cardiovascular investigations.

    Behavior toxicology and psychophysiology and ergonomics.

Publications: In-house publications include:

    <u>Poisoning by Pesticides</u>.

    <u>Agricultural Health</u>.

    <u>A Guide to Respiratory Protection</u>.

    <u>Toxicology of Materials Used in Electroplating</u>.

    <u>Occupational Skin Disease and Tinea Infection</u>.

    <u>Dermatitis Caused by Epoxy Resins</u>.

    <u>Free Services to Industry</u>.

    <u>Inorganic Lead Poisoning</u>.

    <u>Lift the Right Way—Group Lifting</u>.

    <u>Safe Materials Handling</u>.

    <u>Kinetics for Nurses</u>.

    <u>Dust Control</u>.

    <u>Working with Mineral Oil</u>.

Labor

Vehicle Builders Employees Federation of Australia
70 Drummon Street
Carlton, South Victoria
3199 Australia
Telephone: (02) 347-2866; Telex: 30705

Contact:
   Anna Stewart, Research Director

Objectives/Goals:
   To collect information on occupational safety and health.

   To assist workers in both Australia and neighboring Southeast Asian countries to protect themselves against the health and safety hazards generated by multinational companies.

   To link up with infant groups specializing in this area.

Research/Projects:
   Information available upon request.

Publications:
   List available upon request.

CENTRAL AMERICA

Costa Rica

Public Interest

Amigos de la Naturaleza (Inter-American Association for the Study and Defense of the Human Environment)
P. O. Box 162
Guadalupe, Costa Rica

Objectives/Goals:
   To promote and increase individual and collective awareness of environmental concern for human survival.

Research/Projects:
> To promote conservation of flora and fauna and special ecosystems.
>
> To compile and disseminate scientific knowledge on pollution and abuse of the environment.

Publications:
> List available upon request.

Cuba

Government

Instituto Nacional de Medicina del Trabajo
Havana, Cuba

Contact:
> Antonio Granda Ibana, M.D., Director

Objectives/Goals:
> Research on occupational disease prevention, recognition, and treatment.
>
> Educational training of health professionals in occupational health.

Research/Projects:
> Information available upon request.

Publications:
> Information available upon request.

EUROPE

Belgium

Academic

University of Louvain
Department of Occupational Medicine and Hygiene

clos Chapelle-aux-Champs 30
Box 30.54
1200 Brussels, Belgium
Telephone: 02/762-34.00, ext. 32.20

Contact:
    Professor R. Lauwerys, Director

Objectives/Goals:
    Research.

    Teaching.

    Specialized services in occupational medicine and hygiene.

Research/Projects:
    Biological monitoring of workers exposed to industrial chemicals.

    Heavy metals (Pb, Hg, Cd) toxicity.

    Industrial dermatology (mechanism of contact dermatitis).

    Heat stress.

    Noise.

Publications:
    Book on occupational diseases (French, Italian, and Finnish editions).

    Publications list available upon request.

International

European Environmental Bureau
31, Rue Vautier
1040 Brussels, Belgium
Telephone: 647-01-99

Contact:
    G. Verbrugge, Counselor

Objectives/Goals:
> Liaison between 50 environmental organizations in the nine member-states and the European institutions.

Research/Projects: Current research topics include:
> Toxic substances.
>
> Export of toxic products and hazardous technologies.
>
> Pollution control.
>
> Energy.

Publications:
> List available upon request.

Bulgaria

International

Institute of Hygiene and Occupational Health
Boul Dim Nestorov 15
Sofia 1431, Bulgaria
Telephone: 59-10-06; Telex: 22712 MAREKT BG

Contact:
> Professor F. Kaloyanova, M.D., Ph.D., D.SC.,
> Director of Institute

Objectives/Goals:
> The institute is a collaborating center of the World Health Organization, set up to conduct research, training, and therapeutics in the fields of hygiene and environmental and occupational health.

Research/Projects:
> Responsibility for the elaboration of sanitary standards and requirements.
>
> Analyses and prognosis concerning the influence of environment on public health.
>
> Promotion of sanitary control.

Publications:
>Annual publication of the institute concerning problems of hygiene.

>Monograph on the basis of toxicology.

>Monograph on the novelties in toxicology.

>Monograph on the hygiene conditions in the principal branches of industry.

>Monograph on the composition of drinking waters in Bulgaria.

Denmark

Public Interest

International Youth Federation for Environmental Studies and Conservation (IYF)
Klostermolle
Klostermollevej 46
DK-8660 Skanderborg
Denmark
Telephone: 05-782044

Contact:
>Marc Pallemaerts, President

Objectives/Goals:
>To promote and coordinate the activities of voluntary youth environmental groups all over the world, through information on exchange, network building, training courses, international seminars, financial and material support, and joint projects on issues of common concern.

Research/Projects:
>A current project in the field of international environmental and occupational safety and health care in cooperation with the youth forum of the European Economic Community (EEC), which represents all organized youth in the EEC, is to obtain a ban on the exports to developing countries of hazardous pesticides banned within the EEC.

Publications:
> A brochure compiling evidence of the environmental and health effects of pesticides exported from the EEC is being prepared to be used in lobbying the European Parliament on this issue.

Finland

Academic

Helsinki University
Department of Environmental Conservation
SF-00710
Helsinki 71, Finland
Telephone: 90-378011

Contact:
> Pekka Nuorteva, Professor, Department Head

Objectives/Goals:
> Teaching.
>
> Scientific investigations on environmental problems.

Research/Projects:
> Mercury bioaccumulation in nature, especially the role of sarcosaprophagous insects in the bioaccumulation processes.
>
> Occurrence of mercury in mushrooms, lichens, fish, and aquatic plants in the vicinity of caustic soda factories in Finland and Thailand.
>
> Mercury in fish and human hair by the artificial lake of Lokka in the Subarctic.
>
> City garbage dumps as source of heavy metal pollution.
>
> The significance of nitrogenous fertilizers in the formation of carcinogenic N-nitrosoamines.
>
> The significance of summer cottage for the environment and man.
>
> DEHP bioaccumulation in Finnish nature (past).

Publications:
: List available upon request.

Government

Institute of Occupational Health
Haartmaninkatu 1
SF-00290
Helsinki 29, Finland

Contact:
: Sven Hernberg, M.D., Scientific Director

Objectives/Goals:
: To provide education, training, and research activities for the study of occupational safety and health.

Research/Projects:
: Occupational diseases and other illnesses induced by work.

  Occupational accidents and their prevention.

  Industrial hygiene.

  Toxicology.

  Ergonomics.

  Industrial safety problems.

  Labor protection and occupational health care.

  To provide research for practical solutions in occupational health care, labor protection, and standard setting.

Publications:
: Weekly Bulletin, issued to employees 24 times throughout the year.

  Institute Newsletter, quarterly publication in Finnish and English.

  Current Listings of Research Projects, in Finnish and English.

Institute Annual Report, in Finnish and English.

Institute Audiovisual Presentations in Finnish, English, German, Russian, and Swedish.

Additional listings available upon request.

National Board of Health
Health Directorate
Siltasaarenkatu 18 A
SF-00530
Helsinki 53, Finland
Telephone: 718511; Telex: 12-1774

Contact:
    Markku Murtomaa, M.D., Director, Occupational Health Services

Objectives/Goals:
    To guide, supervise, and develop the following nationwide health care services: production and distribution of drugs; supervision of institutions and laboratories; research, planning, standardization, and consulting activities in the health field.

Research/Projects:
    Guidelines, studies, and statistical analyses in occupational health.

Publications:
    Yearbook of The National Board of Health: 1977-1978 (Health Services, 1979).

    Official Statistics of Finland 11, no. 75 (Helsinki, 1979).

France

Public Interest

Collectif Securite; Universites Paris VI et VII
Batiment H
4 Place Jussieu
75230 Paris Cedex 05, France

Contact:
> A. Lascoux

Objectives/Goals:
> To control the use of asbestos.

Research/Projects:
> To expand organization's work to occupational health hazards.

Publications:
> <u>Danger Amiante</u>, 424 pp (Maspero, 1977), $6.00 plus postage.
>
> <u>En finir avec l'Amiante</u>, periodical.
>
> Leaflets on asbestos, technical information.

Germany

Academic

Institut für Hygiene und Arbeitsmedizin
Universitätsklinikum der Gesamt Hochschule Essen
Hufelandstrabe 55-4300 Essen 1
0201 Germany (F.R.G.)
Telephone: 7232570/857, klies d.

Contact:
> Professor Dr. Med. J. Bruch, Department Head

Objectives/Goals:
> Research and teaching in industrial hygiene and medicine.

Research/Projects:
> Research on dust effects and pneumonconiosis.

Publications:
> Available upon request.

Government

Bernhard-nocht—Institute for Nautical and Tropical Diseases, Department for Nautical Medicine

26 / SAFETY AND HEALTH CATALOGUE

Seewartenstrasse 9a
2000 Hamburg, Germany
Telephone: 31.102.490/1

Contact:
    Priv. Doz. Dr. H. Goethe, Scientific Director

Objectives/Goals:
    To carry out scientific work in the field of nautical medicine and cooperative work with the shipbuilding industry and the corresponding departments of governmental institutes for navigation and medicine.

    The department is also official cooperator of the World Health Organization, International Labor Organization, and IMCO and maintains a reference library of nautical medicine that contains the relevant historical and current publications.

Research/Projects:
    Past Projects: Investigations on noise, ergonomics, water supply, psychophysical load on board ships.

    On-going Projects: Investigations on vibration, ergophthalmological problems, physical load and catecholamine excretion among sea pilots, problems of survival at sea.

    Prospective Projects: Continuation of the above problems and studies on watchkeeping problems.

Publications:
    <u>Bibliography on Nautical Medicine</u>, Vol. I (Hamburg, 1977).

    Further publications list available upon request.

<u>Labor</u>

Hauptverband der Gewerblichen
Berufsgenossenschaften
Gesetzliche Unfallversiche-rung
Langwartweg 103
D-5300
Bonn 1, Germany
Telephone: 02221-5491; Telex: 886 628 bavbd d

Contact:
> Dr. Friedrich Watermann, Hauptgeschaftsfuhrer (Director)

Objectives/Goals:
> Accident prevention.

Research/Projects:
> Workplace medicine.
>
> Protection against medicine.
>
> Noise abatement.
>
> Dangerous work substances.
>
> Accident statistics.
>
> Rehabilitation.

Publications:
> Zeitschrift, <u>Die Berufsgenossenschaft</u> (Journal, <u>The Professional Trade Union</u>).
>
> Unfallvarschriften, Sicherheitsregelin (writings on accidents and safety rules).

Public Interest

The Conservation Foundation
Koenigstr. 12 B
D-5300 Bonn 1, Germany
Telephone: (49)-(2221)-22-45-80

Contact:
> Cynthia Whitehead, Associate; European Representative

Objectives/Goals:
> Nonprofit research and communications organization to promote the wise use of the earth's human and natural resources.

Research/Projects:
> Current research includes: toxic chemicals; urban development and the elderly; forestry; industrial siting planning.

Publications:
> Determining Unreasonable Risk Under the Toxic Substances Control Act, Clarence J. Davies, 1979, $4.00.
>
> Chemical Hazard Warnings: Labeling for Effective Communication, Sam Gusman and Frances Irwin, November 1979, $4.00.
>
> Complete list available upon request.

Italy

Academic

Clinica del Lavoro Universita
Instituti Clinici di Perfezionamento
Via San Barnaba 8
20122 Milan, Italy
Telephone: 54-63-707

Contact:
> Professor Antonio Grieco, Director of Clinic

Objectives/Goals:
> Teaching; scientific research and medical care on occupational health.

Research/Projects:
> List available upon request.

Publications:
> List available upon request.

Government

Laboratorio di Igiene del Lavoro, Instituto Superiore di Sanita
Viale Regina Elena 299
00161 Rome, Italy
Telephone: 06-4990; Telex: RM071 ISTISAN

Contact:
> Dr. Marco Biocca

Objectives/Goals: Established in 1976:
  To study human effects of environmental risks.

  To study methods for monitoring environmental exposures and biological indicators.

  To propose criteria for recommended environmental and occupational health standards.

Research/Projects:
  Analyses of chloroorganic chemicals, in particular tetra chlorodibenzodioxins.

  Analyses of organic solvents in workplaces.

  Epidemiological, methodological, and experimental studies on chemical risks existing in specific workplaces.

  Toxicological studies on polyneuropathy in shoe factories.

  Studies on molecular forms of hematic cholinesterases with reference to occupational exposures.

  Studies on asbestos exposures in shipyards and other workplaces.

  Epidemiological studies on dockers and glass workers.

  Studies for recommended criteria for analyses of dusts, in particular silica.

Publications:
  List available upon request.

Health

Coordinamento Nazionale dei Servizi di Medicina del Lavoro
Via Tiarini 22/2
Bologna, Italy

Contact:
  Dr. Leopoldo Magelli

30 / SAFETY AND HEALTH CATALOGUE

Objectives/Goals:
> National association of local Occupational Health Services Personnel. The association was formed by occupational health physicians to study the principal methodological and technical problems in occupational health services.

Research/Projects:
> Information available upon request.

Publications:
> List available upon request.

Labor

Centro Ricerche e Documentazione sui Rischi e Danni da Lavoro
Viale Regina Margherita 37
00100 Rome, Italy

Contact:
> Mr. Gastone Marri

Objectives/Goals:
> Established in 1972 by the trade unions CGIL, CISL, and UIL. They are collecting the most complete information about workers' initiatives in occupational health.

Research/Projects:
> List available upon request.

Publications:
> List available upon request.

Public Interest

Women's International Information and Communication Service (ISIS)
Via delle Pelliccia 31
00153 Rome, Italy
Telephone: 06/580 82 31

Objectives/Goals:
> International Women's Information and Communication Service based in Rome and Geneva

Research/Projects:
> Information available upon request.

Publications:
> Information available upon request.

Luxembourg

Introduction

Commission of the European Communities
Batiment Jean Monet
Avenue Alcide de Gasperi
Luxembourg-Kirchberg
Grand Duchy of Luxembourg
Telephone: 43011; Telex: Comeur Lu 3423

Contact:
> Alexandre Berlin, Head of Specialized Service; W. J. Hunter, Principal Administrator

Objectives/Goals:
> To implement legislation on safety and health at work within the European communities.

Research/Projects:
> Projects and studies on all aspects of health and safety at work.

Publications:
> Publications include conference proceedings, scientific and technical reports and recommendations, documents published by members of the Health and Safety Directorate.

The Netherlands

Government

Netherlands Institute for Preventive Health Care (NIPG) TNO
Wassenaarseweg 56
2333 AL Leiden
P. O. Box 124
2300 AC Leiden, The Netherlands
Telephone: 071-150940

Contact:
> M. J. Hartgerink, Director

Objectives/Goals:
> Research directed to the improvement of preventive health care on medical, psychological, and social issues.
>
> Research on the organizational conditions for optimum health care delivery.

Research/Projects: Current research topics include:
> General occupational and industrial medicine.
>
> Ergonomics and experimental psychological and physiological research.
>
> Child and adolescent preventive health care.
>
> Postgraduate training courses in social medicine and epidemiology include: occupational and industrial medicine; public health administration; seminars on environmental problems; basic courses in public health; child health care.

Publications:
> NIPG/TNO Annual Report.
>
> List available upon request.

Norway

Government

Institute of Occupational Health (Yrheshygienik Institutt)
Gydas vei 8
Oslo 3, Norway

Contact:
> Jor Norseth, M.D., Ph.D., Director

Objectives/Goals:
> To work for the best possible adjustment of the working environment for the worker.

Research/Projects:
    General occupational health and industrial medicine.

    Industrial hygiene, industrial and experimental toxicology.

    Clinical industrial medicine and epidemiology.

Publications:
    Yearly Report (in Norwegian).

    Publications in international journals (publications list found in the yearly report).

Labor

Norwegian Federation of Trade Unions
Landsor Ganisas Onen 1 Norge-LO
Youngsgata 11
Oslo 1, Norway
Telephone: 40-16-96

Contact:
    Borbe Pettersen, Head of Department of Environment

Objectives/Goals:
    To create a better workplace environment.

    To help workers understand the links between environment and health policy.

Research/Projects:
    None.

Publications:
    None.

Public Interest

Natur Og Ungdom
Stenersgata 16
Oslo 1, Norway
Telephone: 02-41-93-11

Objectives/Goals:
>   To achieve a low-energy society based on self-sufficiency and nongrowth.

Research/Projects:
>   Current projects include: traffic and transportation studies; energy (hydropower, nuclear power) studies; oil exploration research.

Publications:
>   <u>Natur Of Samfunn,</u> external magazine with special publications on the following topics, all in Norwegian: energy, agriculture, forestry, transportation, birds.

Romania

<u>Academic</u>

Institute of Hygiene and Public Health
Str. Dr. Leonte Nr. 1-3, Code: 76256
Bucharest, Romania
Telephone: 49-40-30

Contact:
>   Bernard Barhad, Deputy Director

Objectives/Goals:
>   Teaching graduate medical students.
>
>   Teaching postgraduate students.
>
>   Patient care and research.

Research/Projects:
>   Epidemiological investigations on respiratory diseases in engineering factories.
>
>   Occupational diseases of CS.
>
>   Occupational asthma in different workplaces (clinical and epidemiological studies).

Publications:
>   <u>Igiena</u>, hygiene periodical.

Medicina Muncii (textbook), edited by Petru Manu
(Bucharest: Editura Medicala, 1975).

Practica Medicinei Muncii (textbook), edited by Petru
Manu and Toma Niculescu (Bucharest: Editura Medicala,
1978).

Sweden

Government

Arbetarskyddsstyrelsen
(National Board of Occupational Safety and Health)
Ekelundsvagen 16, Solna
S-171 84 Solna, Sweden
Telephone: 46-8-730-90 00

Objectives/Goals:
    To lead, coordinate, and supervise the activities in the field of occupational safety and health.

    To control the observance of the legislation concerning the working hours and working environment.

    To issue regulations on the application of the legislation.

    To provide information and training in occupational safety and health care.

    To provide research, training, and contractual work for technicians, doctors, and nurses.

    To control the observance of the act of products hazardous to health and to the environment and related provisions.

Research/Projects:
    Catalogue of research projects in progress at the Occupational Health Department of the board. Published annually, available in English.

Publications:
    "Newsletter" in English concerning the board's current activities, including annual list of publications and duplicated reports from the board's Occupational Health Department. Four issues a year.

36 / SAFETY AND HEALTH CATALOGUE

Scientific series on occupational health "Arbete Och Hälsa." At least monthly. Available in Swedish with a summary in English.

Annual Report. Available in Swedish with tables in English.

Arbetarskydd. A monthly journal in Swedish.

Reports on Methods ("Metodsrapporter"); Investigations ("Undersoknings-Rapporter"); Training and Education ("Utbildning Srapporter"); and Commissioned Research ("Uppdragsrapporter"). Available in Swedish.

Arbetslivscentrum (Swedish Worklife Center)
Box 5606
Stockholm S-11486, Sweden
Telephone: (08) 229980

Contact:
Birger Viklund, Information Director

Objectives/Goals:
Democratization of worklife through research, education, and information.

Research/Projects:
40 projects primarily in industrial relations research and the effects of the Safety and Health Act.

Study of hazardous environments: "Social Clause (in the GATT), Effects on Developing Countries, Alternatives."

Publications:
Research Reports (Swedish, summaries in English).

Working papers (some in English).

Newsletter.

Documentation and Information Service (25,000 entries in the computer base).

Labor

Joint Industrial Safety Council
Box 3208, 103 64
Stockholm, Sweden
Telephone: 08-22 94 20

Contact:
 Ingvvar Soderstrom, Director

Objectives/Goals:
 Education, information, and consultative activities concerning working environment questions.

Research/Projects:
 Training Materials.

Publications:
 A Better Working Environment—a basic course.

 Further Training Materials: lighting; planning; chemical health hazards; ergonomics; noise; local safety work.

 Research and Development: noise from pumping equipment (suggestion to solutions from the Pulp and Paper Industry); external noise problems in industry; infrasonics in working environment; constructions and material for noise absorption; microwave ovens; radiation protection in industry.

 Laws and Agreements: Swedish legislation on the working environment (overhead visual material on the contents of the legislation); fire protection law; act on products hazardous to health and the environment; external environment protection legislation; working environment agreement (in English).

 Handbooks: Industrial Accidents—a method for internal reports; set of forms to classify illness absenteeism; List of Words for the Working Environment (Swedish-English); Road Safety and Working Environment (instructions for managing these questions in the companies); Fight Against Noise—Principles and Application.

 Research Reports: Cooperation between the Joint Industrial Safety Council and the Swedish Work Environment Fund

38 / SAFETY AND HEALTH CATALOGUE

in order to publish and spread the reports to the working life people; Shift Work and Well-Being; What Are Psycho-Social Working Environment Questions?; Safety Work in Big and Medium-Sized Companies; Working Conditions for Programming Staff; Safety Work within Small Industrial Companies; Training in Better Working Environment; a short version of an evaluation of the basic course, A Better Working Environment.

Switzerland

Academic

Institut Unversitaire de Medecine du Travail et D'Hygiene Industrielle
Route de la Clochatte
CH 1052 Le Mont-sur-Lausanne
Switzerland
Telephone: (021) 322606

Contact:
    Professor M. Guillemin, Director

Objectives/Goals:
    Prevention of occupational diseases through technical, medical, and scientific assistance; consultations; and educational programs.

Research/Projects:
    Improvement of biological monitoring for toxic metals and solvents.

    Analytical and epidemiological studies on exposed workers and the general population.

Publications:
    Internal reports and publications in scientific journals.

Government

The Service of Occupational Medicine of the Federal Office for Industry and Labor
Arbeitsarztlicher Dienst des BIGA

Bundesgasse 8
Bern CH 3003, Switzerland
Telephone: Bern: 61 29 10

Contact:
    Dr. W. F. Greuter, Chief; Dr. J. Buchberger;
    Dr. S. Kundig; Dr. A. Deuber

Objectives/Goals:
    Medical supervision related to the Labor Act; protection of health at work; environmental control inside and outside the workplace.

Research/Projects:
    Occupational risks of noise, heat, exhaust gases, and solvents.

    Occupational risks on account of: styrene exposure in polyester industry; skin exposure to oils and coolants in industry; work on visual display units; shift and night work; exposure to anaesthetic gases (operating theaters); department store/shop assistant work; forest work; identification of occupational risks by man monitoring.

Publications:
    A number of articles in various journals are available upon request.

International

International Register of Potentially Toxic Chemicals (IRPTC)
United Nations Environment Program (UNEP)
W.H.O Building, Room L31
1211 Geneva 27, Switzerland
Telephone: 91-21-11

Contact:
    Dr. J. W. Huismans, Director

Objectives/Goals:
    To collect and publish data on the hazardous effects of chemical substances. (See Appendix J for a more complete description of IRPTC and UNEP.)

Research/Projects:
To identify the characteristics of potentially hazardous chemicals.

To define hazardous characteristics and programs in the data bank.

Publications:
IRPTC attributes for a chemical data register.

Data profiles of chemicals for the evaluation of their hazards to the environment of the Mediterranean Sea.

World Health Organization
Health Legislation
1211 Geneva 27, Switzerland
Telephone: (022) 912457

Contact:
S. S. Fluss, Senior Editor, International Digest of Health Legislation

Objectives/Goals:
To monitor and transfer information on national and international health legislation in all fields, including occupational health.

Research/Projects:
Comparative survey of legislation on protection against ionizing radiation, published 1971.

Publications:
International Digest of Health Legislation (quarterly in English and French).

Labor

International Metal Workers' Federation
Health and Safety Department
54 Bis Route des Acacias
1227 Geneva, Switzerland
Telephone: (022) 43-61-50; Telex: 23298

Contact:
    Rolf Ahlberg, Director

Objectives/Goals:
    To help and support democratically constituted trade unions on different union questions.

Research/Projects:
    To educate trade union representatives in health and safety matters in Africa, Asia, and Latin America. First stage of three-year project is training instructors. Second stage is training safety delegates and shop stewards.

Publications:
    Educational publication for safety delegates: <u>Health Hazards with Solvents, Paints, and Plastics</u>, available in English, French, German, Spanish, and Swedish.

    <u>Health Hazards with Metals</u>, available in English and Spanish.

    <u>Check-list for Safety Delegates</u>, available in English, Spanish, Portuguese, and Swedish.

    Educational program with four films, available in English, Spanish, and Portuguese.

    <u>IMF Health and Safety Bulletin</u>, in English, French, German, Spanish, and Swedish.

United Kingdom

<u>Academic</u>

University of Aston in Birmingham
Department of Safety and Hygiene
Gosta Green
Birmingham B4 7ET
United Kingdom
Telephone: 021-359-3611, extension 6342

Contact:
    Mrs. S. V. Kendell, Information Coordinator

Objectives/Goals:
> To develop occupational safety and health knowledge through research.
>
> To coordinate safety and health knowledge into a coherent discipline.
>
> To provide training and research facilities for students interested in the field of occupational safety and health.
>
> To solve industrial safety and health problems.

Research/Projects:
> Current research topics include: chemical process safety; economic aspects of accident prevention; health and safety legislation; machinery guarding; manual handling; noise and vibration; occupational accidents, theory, and analysis; occupational cancer; occupational pathology; occupational risk prevention; theory and organization; personal protection equipment; rescue and emergency procedures; toxic and dangerous substances; ventilation and air conditioning.

Publications:
> List available upon request.

University of Manchester
Department of Occupational Health
Stopford Building
Oxford Road
Manchester M13 9PT
United Kingdom
Telephone: (061) 273-8241/Manumen Manchester

Contact:
> C. J. Whitaker, Research Associate

Objectives/Goals:
> To conduct research in occupational health and hygiene.
>
> To advise industries on health problems.
>
> To act as medical consultants in cases of occupationally related diseases.

Research/Projects:
> The role of microorganisms in the aetiology of byssinosis.
>
> The relationship between biochemical changes and neurophysiological findings in lead poisoning.
>
> The respiratory effects of enzyme inhalation.
>
> Occupational allergens, in particular, ocamylase and papain.
>
> Distance teaching of occupational health to doctors. Occupation as a precursor of cancer.

Publications:
> Department has available copies of articles published in learned journals on the above research topics.

Government

Health and Safety Executive
Baynards House
1 Chepstow Place
London W2 4TF
United Kingdom
Telephone: 01-229-3456

Contact:
> John Locke, Director

Objectives/Goals:
> To enforce national statutory requirements on health and safety.

Research/Projects:
> Extensive list available upon request.

Publications:
> Extensive list available upon request.

Health

Guy's Hospital Medical School
Department of Community Medicine

Unit for the Study of Health Policy (USHP)
8, New Comen Street
London, SE1 1YR
United Kingdom
Telephone: 01-407-7600, ext. 2999

Contact:
    Dr. Peter Draper, Director

Objectives/Goals:
    To promote the informed public discussion of issues of health policy.

    To produce reports based on published research, official statistics, and statements of government policy.

    To contribute to the development of a contemporary Public Health Movement for Social Change that is health-promoting rather than health-damaging.

Research/Projects:
    Unit reports, papers, and talks have stressed three themes: The conflicts between conventional ideas of economic growth and health; the image of "progress in health" fostered by the mass media; the demise of public health.

    Future research projects include: The problems of securing changes that are sound in terms of human ecology; improving the discussion of issues of health policy in the mass media; resource allocation and other organizational issues in the health service; social indicators with special relevance to health and the inadequacy of economic indicators as substitutes for social indicators; academic lessons to be learned from the interdisciplinary study of health policy.

Publications:
    <u>Health, Money and the National Health Service</u>, April 1976.

    <u>Economic Policy and Health</u>, November 1976.

    <u>Health, the Mass Media and the National Health Service</u>, 1977.

    <u>The NHS in the Next 30 Years: A New Perspective on the Health of the British</u>, July 1978.

Rethinking Community Medicine: Towards a Renaissance in Public Health? June 1979.

Health and Safety at Work Magazine
Maclaren Publishers Ltd.
Davis House 69 High Street
Croydon CR9 1QH
Surrey, England
Telephone: 01-688-7788, Ext. 58; Telex: 946665

Contact:
   John Manos, Deputy Editor

Objectives/Goals:
   Publication of news and feature articles on current developments in legal, medical, industrial hygiene, industrial relations, and other aspects of health and safety at work.

Research/Projects:
   Information available upon request.

Publications:
   Information available upon request.

Labor

Association of Scientific, Technical, and Managerial Staffs (ASTMS)
Whitehall Office
Dane O'Coys Road
Bishops Stortford, Herts
United Kingdom
Telephone: 0279-58111

Contact:
   Sheila McKechnie, Health and Safety Officer

Objectives/Goals:
   White-collar trade union activities.

Research/Projects:
   None, at present.

Publications:
> Policy documents: Guide to Health and Safety at Work; Guide to Health Hazards of Visual Display Units; Prevention of Occupational Cancer, February 1980.

General and Municipal Workers Union (GMWU)
Thorne House
Ruxley Ridge, Claygate
Esher, Surrey
United Kingdom
Telephone: ESHER 62081; Telex: 27428

Contact:
> David Gee, Health and Safety Officer

Objectives/Goals:
> To improve occupational health care programs in the workplace.

Research/Projects:
> Seminars on criteria for carcinogenicity in the workplace.
>
> Cancer prevention programs.
>
> Research projects on: radiation; shiftworking; pesticides; noise; dust and ventilation.

Publications:
> Safety Representatives Handbook.
>
> HAZARD.
>
> Model letter and safety data sheets from union safety representatives.

Public Interest

British Society for Social Responsibility in Science (BSSRS)
9 Poland Street
London W1V 3DG
United Kingdom
Telephone: 01-437-2728

Contact:
> Alan Dalton, Jr., BSSRS Work Hazards Group

Objectives/Goals:
> Established in 1969 by a group of socially concerned scientists to understand and publicize how scientific developments affect people at work. The BSSRS also supports workers and community groups in their struggles for healthy and safe working conditions.

Research/Projects:
> Provides education and training of safety representatives, shop stewards, and other workers concerning the hazards of work, usually on day-release safety training courses organized by the Trade Union Congress and WEA.
>
> Provides scientific information in direct response to queries from workers and their organizations.
>
> Encourages the organization of: workplace trade union health and safety committees, whether or not joint consultative safety committees already exist; area health and safety organizations of workers, tenants, and residents either as area committees like the Coventry Health and Safety Movement or as part of the trades council; effective channels within each trade union to provide the information and support needed by its members; aims to help workers to challenge management standards and to set and control their own standards, putting health and safety first.

Publications:
> Hazards Bulletin, published five times a year to provide information on specific hazards, legal cases, and related publications.
>
> Pamphlets on specific hazards including: Noise, 1975, 1977; Oil (mainly in machine shops), 1976; Vibration, 1977; Asbestos, 1979.
>
> List available upon request.

BSSRS Work Hazards Group: Scientists and trade unionists who produce a national Hazard Bulletin and various booklets or leaflets and provide advice and speakers on occupational safety and health concerns. The current groups include:

48 / SAFETY AND HEALTH CATALOGUE

Birmingham
Saltley Action Centre
2 Alum Rock Rd.
Birmingham (021-328-4184)

Brighton
Ian Wright
68 Compton Rd.
Brighton, Sussex

Bristol
85 Ashley Road, Montpelier
Bristol BS6 5 NR (STD 0272-55 4660)

Hospital Hazards Group
Gene Feder, BSSRS
9 Poland Street
London WIV 3DG (01-437-2728)

London
Alan Dalton, London Work Hazards Group,
BSSRS
9 Poland Street
London WIV 3DG (01-437-2728)

Manchester
Ken Green
9 Dalston Drive
Manchester M20 00L (061-445-1096)

Merseyside
Mary Crimmons
70 Granville Road
Liverpool 15 (051-733-6925)

Northeast
13 Railway Street
Langley Park, Durham (0385-731889)

Sheffield
Dave Hayes
14 Goodwin Road
Sheffield 8 (0724-57337/583856)

Women and Work Hazards Group
Marianne Craig, BSSRS

9 Poland Street
London WIV 3DG (01-437-2728)

Local Trade Union Health and Safety Groups: Organized to assist working people over health and safety issues. They conduct advice centers, workshops, and lists of related publications.

Bedford—BASH
Bedford Association for Safety and Health
146 Spring Road
Kempston, Bedford (0234-66755)

Birmingham—BRUSH
Birmingham Regional Union Safety and Health Campaign
160 Corisande Road
Selly Oak, Birmingham 29

Bristol—BASH
Bristol Action on Safety and Health
6 Keynes Road
Clevedon, Bristol

Cannock—CHASE
Cannock Health and Safety Experiment
56 Orange Crescent
Penkridge, Staffs

Coventry—CHASM
Coventry Health and Safety Movement
229 Bredon Avenue
Binley, Coventry (0203-456635)

Doncaster—HASSARD
29 High Street
Arksy, Doncaster DN5 0SF

Dunfermline—DASH
Dunfermline Area Safety and Health Group
Clackmannon House
Clackmannon

Leeds—LASH
Leeds Action on Safety and Health
29 Blenheim Terrace
Leeds 2 (0532-39633)

50 / SAFETY AND HEALTH CATALOGUE

London North
North London Health and Safety Group
c/o Camden Community Law Centre
146 Kentish Town Road
London NW1 (01-485-6672)

London South—SLASH
South London Action on Safety and Health
c/o 506 Brixton Road
London SW9 (01 723-4245)

London South-East—HASSEL
Health and Safety in South East London
6 Sedgebrook Road
London SE3

London West—MASH
Middlesex Action on Safety and Health
c/o Hillingdon Community Law Centre
63 Station Road
Hayes, Middlesex (01-573-4021)

Manchester—MASC
Manchester Area Safety Committee
4 Montrose Avenue, West
Didsbury, Manchester M20 BLN

Merseyside Hazards Groups—MHG
Merseyside Hazards Group
70 Granville Road
Liverpool 15 (051-733-6925)

Rotherham—WEA
WEA Health and Safety Information Service
Chantry Buildings
Corporation Street
Rotherham (0709-72121)

Sheffield—TUSC
Sheffield Trade Union Safety Committee
312 Albert Road
Sheffield 8 (0742-584559)

Southampton—WHAC
Work Hazards Advisory Committee

27 Pointout Road
Bassett, Southampton

Telford—THSG
Telford Health and Safety Group
67 High Street
Dawley, Telford (0952-501-484)

Earth Resources Research Ltd. (ERR Ltd.)
40 James Street
London, W1
United Kingdom
Telephone: (01) 487-4835

Contact:
  David Baldock

Objectives/Goals:
  Independent environmental research and policy group.

Research/Projects:
  Assessment on the impact of the banking policies of the European Investment Bank.

  Survey of agribusiness in black Africa.

  Unemployment caused by new microelectronic technologies in the United Kingdom.

Publications:
  <u>Automatic Unemployment</u>, ERR 1979, C. Aimes and G. Seoule.

Friends of the Earth Limited (FOE)
9 Poland Street
London W1V 3DG
United Kingdom
Telephone: 01-434-1684

Contact:
  Tom Burke, Director, Special Projects

Objectives/Goals:
> The achievement of an equitable and sustainable society in which economic and social policies are shaped by a full understanding of their environmental context. FOE is committed to the reduction of the environmental impact of human activities and to the development of alternatives to ecologically harmful practices.

Research/Projects:
> Currently mounting five major campaigns in: wildlife; land use; resources and bicycles; transportation; energy.

Publications:
> List available upon request.

International Institute for Environment and Development (IIED)
27 Mortimer Street
London W1N 8 DE
United Kingdom
Telephone: 01-580-7656-7

Contact:
> Barbara Ward, President

U.S. office:
1302 Eighteenth Street, N.W.
Suite 501
Washington, D.C. 20036
(202) 462-0900

Objectives/Goals:
> Nonprofit organization that serves as a catalyst for information action on environmental and developmental problems in the international community.

Research/Projects:
> Dissemination of ideas to governments and international agencies to assist their formulation of policies at regional and global levels.
>
> Organization of expert consultant groups that focus attention on specific problems and suggest their solutions as well as methods to achieve them.

Arousal of both public and nongovernmental organizations' interest in the policies needed to solve particular world problems and assisting such groups to organize their efforts around these issues.

Earthscan, an independent environmental information service whose task is to make the world's media aware of important issues centered around environment and development.

Publications:
List available upon request.

National Occupational Hygiene Service, Ltd.
12 Brook Road
Fallowfield
Manchester M14 6UH
United Kingdom
Telephone: 061-224-2332/3

Contact:
Edward King, Chief Executive

Objectives/Goals:
Private, nonprofit company to provide onsite and laboratory services on various aspects of occupational hygiene to all on a simple fee-for-service basis.

Research/Projects:
Dose/Absorption relationships for lead, cadmium, and so on. King et al., "Industrial Lead Absorption," Ann. Occupational Hygiene 22 (1979):213-39.

Publications:
None.

North London Health and Safety Group
c/o Camden Community Law Centre
146 Kentish Town Road
London NW1 9QG
United Kingdom
Telephone: 01.485-6672

54 / SAFETY AND HEALTH CATALOGUE

Contact:
    Gregory Cohn

Objectives/Goals:
    To promote trade union and community interests in health and safety issues in the workplace and in the environment of North London.

Research/Projects:
    Health and Safety Resource Centre: To provide information for the community and specifically for recently appointed safety representatives.

    Enquiry Answering Services: Advice sessions discuss specific health and safety problems, dangers of the hazards, legal rights, and advice on hazard removal.

Publications:
    <u>Information Bulletin</u>, monthly newsletter.

    List of publications available upon request.

OXFAM
274 Banbury Road
Oxford OX2 7DZ
United Kingdom
Telephone: 0865.56777; Telex: 83610

Contact:
    David Bull

OXFAM-America
302 Columbus Avenue
Boston, Massachusetts 02116
Telephone: (617) 247-3304

Objectives/Goals:
    Nonprofit, international development agency that funds self-help programs in Asia, Africa, and Latin America. Economic and food self-reliance are emphasized in all its programs.

Research/Projects:
    Currently OXFAM supports or contributes toward over 1,000 projects designed: for the poorest to have more,

particularly in terms of food, better health, and a fair share of the world's increasingly limited resources; for the poorest to have more in terms of confidence, relationships, self-determination, and ability to manage their own futures.

Publications:
Extensive list available upon request.

Social Audit Limited
9 Poland Street
London WIV 3DG
United Kingdom
Telephone: 01-734-0561

Contact:
Charles Medawar, Director

Objectives/Goals:
Independent, nonprofit organization concerned with improving government and corporate responsiveness to the public.

Publishing arm of the Public Interest Research Centre Ltd., England, a registered charity that conducts research into government and corporate activities.

Research/Projects:
To investigate and report on government and corporate activities. Additional information available upon request.

Publications:
Social Audit Journal.

The Social Audit Pollution Handbook: How to Assess Environmental and Workplace Pollution, Maurice Frankel (London: Macmillan, 1978).

The Social Audit Consumer Handbook: A Guide to the Social Responsibilities of Business to the Consumer, Charles Medawar (London: Macmillan, 1978).

Social Audit: Insult of Injury? An Enquiry into the Marketing and Advertising of British Food and Drug Products in the Third World, Charles Medawar (London: Social Audit Ltd., 1979).

Report of an Enquiry into the Effectiveness of Disclosure of Information on Toxic Hazards to Workpeople, Maurice Frankel (London: Macmillan, 1979).

Additional list available upon request.

The Society for the Prevention of Asbestos and Industrial Disease (SPAID)
38 Drapers Road
Enfield, Middlesex
EN2 8LU United Kingdom
Telephone: (01) 366-1640

Contact:
Nancy Tait, Secretary and Trustee

Objectives/Goals:
Long-term: to prevent industrial disease, especially industrial cancers.

Immediate: to provide an advisory service for workers and to facilitate the exchange of information between workers, researchers, and administrators.

Research/Projects:
Case history studies to identify new risks to workers and the community.

Study of particle fiber size in dusts produced by construction workers using various building materials.

Publications:
Asbestos Kills, Nancy Tait, 1979.

Compensation for Asbestos Diseases.

SPAID NEWS, three issues yearly.

Women and Work Hazards Group
9 Poland Street
London WIV 3DG
United Kingdom
Telephone: (01) 487-2728

Contact:
> Marianne Craig, Coordinator

Objectives/Goals:
> To raise the issue of occupational health in the women's movement.
>
> To research and publish information on the occupational safety and health concerns of women workers.

Research/Projects:
> Teaching occupational safety and health to safety representatives.
>
> Working and consulting with the London Hospital Hazards Group.

Publications:
> Women's Safety and Health Package, 1979.
>
> Health Hazards of Office Work and How to Fight Them.

Yugoslavia

Academic

Institute for Medical Research and Occupational Health
Yugoslav Academy of Sciences and Arts
Mose Pijade 158, pp. 29
41001 Zagreb, Yugoslavia
Telephone: 222600-165

Contact:
> Professor Tihomil Beritic, Head of Clinical Department for Occupational Diseases

Objectives/Goals:
> Medical research.

Publications: Recent projects and publications include:
> "Morphologic Changes of Erythropoietic Cells in the Bone Marrow of Workers Exposed to Lead," Report to the XI International Congress on Occupational Health, T. Beritic and M. Vandekar (Naples, 1954).

"Some Observations of the Morphology of Erythropoietic Cells in Human Lead Poisoning," Blood, T. Beritic and M. Vandekar, 1956.

"Two Cases of Meta Kinitrobenzene Poisoning with Unequal Clinical Responses," Brit. J. Industr. Med., T. Beritic, 1956.

"Iron Containing Blood Cells in Human Lead Poisoning," Report to the XII International Congress on Occupational Health, T. Beritic, Z. Grgic, A. Sirec, 1957.

"Siderotic Granules and the Granules of Punctate Basophilia," Brit. J. Haemol., T. Beritic, 1963.

"Arsenical Polyneuropathy: Poor Effect of BAL upon the Course and Prevention," Report, Brit. J. Haemol., T. Beritic, 1963.

"Compensated and Decompensated Action of Lead upon Human Erythropoiesis," Conference on Inorganic Lead, T. Beritic, 1968.

"Lead Blood Levels and Lead Colic," Conference on Inorganic Lead, T. Beritic, 1968.

"Lead Poisoning from Lead-Glazed Pottery," Lancet, T. Beritic and D. Stahuljak, 1961.

"Blood Lead in Clinical Lead Poisoning," 19th International Congress on Occupational Health, T. Beritic, Telisman Spomenka, Karacic Visnja, Prpic-Majic Danica, Kersanc Antonija, and Pongracic Jadranka, 1978.

"Delta-Aminolevulinic Acid and Coproporphyrin in Urine in the Delayed Action of Lead," XIX International Congress on Occupational Health, T. Beritic, Markicevic Ana, Telisman Spomenka, Karacic Visnja, Pongracic Jadranka, and Kersanc Natonija, 1978.

"An Outbreak of Red Paprika Pepper Poisoning," 8th International Meeting of the European Association of Poison Control Centers, T. Beritic and M. Aljinovic, 1978.

"Lead Concentration Found in Human Blood in Association with Lead Colic," Arch. Environ. Health, T. Beritic, 1971.

"Asbestos and Ferruginous Bodies," Arh. hig. rada, T. Beritic, D. Dimov, Bunarevic Anka, Marija Sondic, and Sirec Anica, 1971.

"ALAD/EP Ratio as a Measure of Lead Toxicity," J. Occup. Med., T. Beritic, Prpic-Majic Danica, Karacic Visnja, and Telisman Spomenka, 1977.

MIDDLE EAST

Egypt

Government

National Institute for Occupational Safety and Health, United States and The High Institute of Public Health, Egypt
(NIOSH-HIPH Occupational Health Research Centre)
University of Alexandria
165 Horria Avenue
Alexandria, Egypt
Telephone: 74895 Alexandria

Contact:
    Dr. Madbuli H. Noweir, Head, Occupational Health Department

Objectives/Goals:
    Teaching, research, and consultation in occupational health.

    Conducts applied research in industry with emphasis on expose-response/effect relationships.

    Carries out investigations of health problems and industrial communities in association with the work and living environment.

    Studies the best feasible and economic control measures that suit the variable situations in industry.

60 / SAFETY AND HEALTH CATALOGUE

> Trains specialists in occupational health and industrial hygiene from Egypt and other countries in the Middle East and Africa.
>
> Provides research facilities for Master and Doctoral candidates in occupational health.

Research/Projects:
> Evaluation of exposure to cotton and flax dust.
>
> Evaluation of exposure to noise in the textile industry.
>
> Gravimetric assessment of silica-containing dust.
>
> Evaluation of exposure to cement.

Publications:
> <u>Bulletin of The High Institute of Public Health</u> (quarterly).
>
> List available upon request.

Israel

Academic

The Hebrew University-Hadassah Medical School
Department of Occupational Health
Jerusalem, Israel
Telephone: 4288493

Contact:
> Professor Marcus Wassermann, M.D., Head of the Department

Objectives/Goals:
> Teaching occupational safety and health and environmental toxicology for graduate students in Public Health and Medicine.

Research/Projects:
> Effects of organochlorine compounds on man.

Publications:
> Reports and papers published in scientific journals.

Kuwait

Academic

University of Kuwait
P. O. Box 24923
Safat, Kuwait
Telex: KUNIVER

Contact:
    Dr. Mustafa Khogali, Faculty of Medicine

Objectives/Goals:
    Teaching, research, and services in international occupational health.

Research/Projects:
    Occupational health in developing countries.

    Occupational pulmonary diseases with emphasis on byssinosis.

    Health problems of migrant workers.

    Heat and heat-induced illness.

    Threshold limit values in developing countries.

Publications:
    List available upon request.

NORTH AMERICA

CANADA

British Columbia

Government

Worker's Compensation Board of British Columbia
5255 Heather Street
Vancouver, B.C. V5Z 3L8
Telephone: (604) 266-0211, Local 447; Telex: 04-507765

Contact:
> A. L. Riegert, Director, Research and Education Department

Objectives/Goals:
> Prevention Services Division promotes the creation of safe work environments through the elimination or control of health and safety hazards and conditions that may lead to industrial injury or disease.
>
> To promote the ability of workers to perform their tasks with no undue risk to themselves or to others in the following ways: inspection of work sites; enforcement of health and safety regulations; education; research; engineering assistance and certification of first aid attendants; blasters and audiometric technicians.

Research/Projects:
> Current research includes: study of red cedar asthma in sawmill workers; health study of pulp mill workers; study of aluminum reduction workers; study of vibration white finger disease of chain saw users; failure of construction cranes; statistical studies of injuries and diseases.

Publications:
> Faller and Bucker Handbook.
>
> Safe Yarding and Loading Handbook.
>
> Construction Safety Handbook.
>
> First Aid Training Manual and First Aid Training Syllabus.
>
> Industrial Audiometric Technician Training Manual.
>
> Lead Poisoning in Industry.
>
> Protecting Your Skin—The Causes and Prevention of Industrial Dermatitis.
>
> Balk Talk.
>
> W.C.B. News—Bimonthly Board Publication.
>
> Workers Compensation Reporter—Volumes 1, 2, 3, 4 available. (These are published decisions of the board in

the areas of claims, rehabilitation, assessment and industrial health and safety.)

Laboratory Analytical Methods—Industrial Hygiene.

Industrial Health and Safety Regulations.

Industrial First Aid Regulations.

Newfoundland

Government

Department of Labour and Manpower, Occupational Safety and Health Division
Province of Newfoundland
Confederation Building
St. John's, Newfoundland AIC 5T7
Telephone: (709) 737-2694

Contact:
    Dr. A. B. Colohan, Executive Director

Objectives/Goals:
    To establish occupational safety and health standards and enforce by legislative authority the act and regulations throughout the Province of Newfoundland.

Research/Projects:
    Current research includes: Asbestos dust in Bale Verte, Newfoundland; effects of silica in Labrador City.

Publications:
    Annual Report.

Nova Scotia

Government

Cape Brenton Development Corporation, Health Services Branch
P. O. Box 2500
Sydney, Nova Scotia PIP 6K3
Telephone: (902) 849-3200

64 / SAFETY AND HEALTH CATALOGUE

Contact:
    Dr. Albert Prossin, Executive Director

Objectives/Goals:
    Promotion and preservation of occupational safety and health.

Research/Projects:
    Current studies include: Effects of coal dust; prevention of pneumoconiosis.

Publications:
    Development of Occupational Health Services in Coal Mining.

    Employee Assistance and Alcoholism Program in Industry.

Ontario

Academic

Queen's University Faculty of Law
Kingston, Ontario K7L 3N6
Telephone: (613) 547-5860

Contact:
    Terence George Ison, Professor of Law

Objectives/Goals:
    Education and research.

Research/Projects:
    Information available upon request.

Publications:
    "The Uses and Limitations of Sanctions in Industrial Health and Safety," 2 Workers' Compensation Reporter 203 (1973).

    The Dimensions of Industrial Disease, Research and Current Issue Series No. 35, Industrial Relations Centre, Queen's University, 1978. ISBN: 0-88886-089-7. Also published in 118 Canadian Medical Association Journal 200 and 317.

    Occupational Health and Wildcat Strikes, Industrial Relations Centre, Queen's University, 1979. ISBN: 0-88886-102-8.

The Forensic Lottery: A Critique on Tort Liability as a System of Personal Injury Compensation (London: Staples Press, 1968).

"Human Disability and Personal Income." A Chapter in Studies in Canadian Tort Law, 2d ed. (Toronto: Butterworths, 1977).

"The Politics of Reform in Personal Injury Compensation," 27 University of Toronto Law Journal 385 (1977).

Information Access and the Workmen's Compensation Board, Research Publication No. 4 of the Commission on Freedom of Information and Individual Privacy, 1979. ISBN: 0-7743-3227-1.

Government

Canadian Centre for Occupational Health and Safety
150 Main Street West
Room 435
Hamilton, Ontario L8P 1H8
Telephone: (416) 523-2611

Contact:
　　Dr. Gordon Atherley, President

Objectives/Goals:
　　To promote health and safety in the workplace and the physical and mental health of working people in Canada.

　　To facilitate: (1) consultation and cooperation among federal, provincial, and territorial jurisdictions; and (2) participation by labor and management in the establishment and maintenance of high standards of occupational health and safety appropriate to the Canadian situation.

　　To assist in the development and maintenance of policies and programs aimed at the reduction or elimination of occupational hazards.

　　To serve as a national center for statistics and other information relating to occupational health and safety.

Research/Projects:
: National Work Injury Statistics Program.

: Hosting 10th World Congress, 1983.

: Sponsored International Symposium on OSH Standards.

Publications:
: None currently.

Department of National Health and Welfare, the Medical Services Branch
Occupational Health Unit
Ottawa, Ontario KIA OL3
Telephone: (613) 996-4921

Contact:
: Dr. T. F. McCarthy, Senior Consultant, Occupational Medicine

Objectives/Goals:
: Provision of health services to Public Service of Canada.

Research/Projects:
: Information available upon request.

Publications:
: List available upon request.

Labor

Canadian Union of Public Employees
233 Gilmour Street
Suite 800
Ottawa, Ontario K2P OP5
Telephone: (613) 237-1590

Contact:
: Colin Lambert, Special Assignments Officer

Objectives/Goals:
: Union activities.

Research/Projects:
: Catalogue of work hazards encountered in the public sector.

Publications:
> Health and Safety Hazards Faced by Canadian Public Employees, Vol. 1.

Ontario Federation of Labour
15 Gervais Drive,
Don Mills, Ontario M3C IY8

Contact:
> Linda Jolley, Coordinator

Objectives/Goals:
> Training in occupational safety and health for all workers in Ontario.
>
> Instructional techniques for trade unionists who teach health and safety.

Research/Projects:
> Workers' Manual on Health and Safety for Ontario and Canada.
>
> Audiovisual presentations on cancer, noise, and effects of toxics on the body.

Publications:
> OFL Occupational Health and Safety Manual ($20.00).
>
> OFL Occupational Health and Safety Instructor's Manual ($15.00).
>
> "At the Source," newsletter published every third month.

Private/Industrial

INCO Limited
1 First Canadian Place
44th Floor
Toronto, Ontario M5X 1C4
Telephone: (416) 361-7511

Contact:
> Dr. E. Mastromatteo, Director, Occupational Health

Objectives/Goals:
>To protect the health of INCO employees.

>To prevent job-related illness.

>To promote positive health practices.

>To develop corporate policies to achieve these goals and to ensure satisfactory performance by the operating units.

Research/Projects:
>Major epidemiological study in cooperation with local unions and directed by McMaster University, Hamilton, Ontario, Canada.

>In-house research on the effectiveness of hearing protection and respiratory protective equipment.

Publications:
>None.

Public Interest

Canadian Environmental Law Research Foundation (CELRF)
5th Floor South
8 York Street
Toronto, Ontario M5J 1R2
Telephone: (416) 366-9717

Contact:
>Michael Perley, Executive Director

Objectives/Goals:
>To promote through legal channels standards and objectives to ensure the maintenance of environmental quality in Ontario and throughout Canada.

Research/Projects: Current projects include:
>Studies of the Environmental Protection Act in Ontario.

>Compensation schemes for pollution victims.

>Handbook on the legal aspects of occupational safety and health.

Publications:
> Two editions of CELRF Handbook, Environment on Trial, on Ontario and federal environmental law.
>
> Two regular case law reporters entitled Canadian Environmental Law Reports and The Canadian Environmental Law Association Newsletter.
>
> Additional publications include a variety of briefs, technical papers, and submissions to various levels of government focusing on the need for law reform in the area of environmental protection.
>
> All publications, audiovisuals, and periodicals available upon request.

Quebec

Academic

Confederation of National Trade Unions (CNTU)
Universite du Quebec a Montreal
Departement des Sciences Biologiques
Montreal, Quebec H3C 3P8
Telephone: (514) 282-7963

Contact:
> Karen Al-Aidroos, Professor, Genetics

Objectives/Goals:
> To undertake projects involving education and research for the CNTU and the Quebec Federation of Labour.

Research/Projects:
> Current research includes: Chromosome studies in male factory workers in a metal refinery—the study will determine damage caused by radiation or numerous other pollutants present in the factory air; preparation of extensive report on women's occupational health and safety problems in Quebec in collaboration with other union women.
>
> Prospective studies: The Health and Textile Workers, jointly supervised by a graduate student and historian; reproduction and VC among male and female workers to be completed by a graduate student in 1981.

70 / SAFETY AND HEALTH CATALOGUE

Publications: All publications are available in French only.
An Exposition on Occupational Safety and Health of Women, by Nicole Vezina and students.

Reproductive Damage to Males and Females and the Relationship to Cancer.

"Work and the Pregnant Woman," by Donna Mergler, published, in part, in Cahiers de la Femme.

Universite du Quebec a Montreal
C.P. 8888 Montreal
Quebec
Telephone: (514) 282-7102

Contact
Donna Mergler, Professor

Objectives/Goals:
Information exchange with the Quebec Foundation of Labour and the Confederation of National Trade Unions on Occupational Health Problems.

Research/Projects: Current work includes:
Occupational health problems of male and female workers in slaughterhouses.

Series of brochures on the health hazards of metals, temperatures, solvents, gases, and dusts.

Occupational safety and health reference center for workers.

Publications:
Noise in the Workplace, brochure covering noise, its measurement, auditory and other physiological effects; noise reduction and legislation.

D. Mergler, F. Ouellet, D. LeBorgue, and S. Simoneau, Le bruit en milieu de travail (Montreal: Instit de recherche applique sur le travail, 1979).

D. Mergler and L. Desnoyers, "La recherche scientifique et la sante au travail," Cahiers du socialisme 3 (1979).

L. Desnoyers and D. Mergler, La crise economique et la sante au travail (Montreal: Colloque la crise et les travailleurs, 1979).

K. Al-Aidroos and D. Mergler, "Les femmes et la sante au travail," Cahiers de la Femme, Toronto, 1979.

Government

Information-Documentation Centre de Toxicologie du Quebec
2705, boul. Laurier,
Ste-Foy, Quebec GIV 462
Telephone: (418) 656-8090, 656-8092 (laboratory)

Contact:
Dr. Albert Nantel, Director; Dr. Jean-Yves Savoie, Head of Laboratory; Dr. Jean-Louis Benedetti, Head, Information-Documentation.

Objectives/Goals:
Responsible for the provincial poison control program and the industrial toxicology program in the province of Quebec: clinical toxicology; industrial toxicology; environmental toxicology; analytical toxicology.

Research/Projects:
The effects of methyl-mercury contamination in the population of the northwest of Quebec.

Human contamination by environmental pollutants near a copper smelter.

International interlaboratory control program in industrial toxicology.

Various clinical and analytical toxicology projects.

Publications:
Prevention of Poisoning in Children, tape-slide show.

Monograph on the surveillance programs of workers exposed to lead, mercury, carbon monoxide, and vinyl chloride (forthcoming).

72 / SAFETY AND HEALTH CATALOGUE

Publication on the diagnosis and treatment of the most frequent poisonings (forthcoming).

Labour Court, Government of Quebec
255 blvd. Cremazie East
7th floor
Montreal, Quebec H2M IL5
Telephone: (514) 873-5543

Contact:
Judge René Beaudry, the Provincial Court

Objectives/Goals:
Entrusted with rendering decisions in labor litigations.

Research/Projects:
Information available upon request.

Publications:
Final Report, October 1976, Quebec Official Publisher (three volumes and eight documents).

Revue Vie Médicale au Canada Francais, Vol. 6, December 1977. "Refléxions sur la prevention des maladies professionelles," p. 1289.

"Les dangers de l'amiante," Encyclopaedia Universalis, 1979, p. 27.

"Le Tribunal du travail vu sous certains aspects," McGill Law Journal, 1974, special edition.

Labor

Quebec Federation of Labour
1290 St. Denis Street
Montreal, Quebec H2X 3J7
Telephone: (514) 288-7431

Contact:
Emile Boudreau, Director, Occupational Health and Safety

Objectives/Goals:
Labor organization.

Publications:
> Le Controle des Travailleurs sur Leur Santé, convention document, December 1975.
>
> La FTQ et Le Livre Blanc, symposium on the Government White Paper, November 1978.
>
> Memoire de la FTQ a la Commission Parlementaire Concernant le Projet de Loi 17, "Loi sur la sante et la securite du travail," (September 1979) (including "Releve des revendications officielles de la FTQ en matiere de sante et securite au travail," January 1975 to November 1978).
>
> Notes, "Verbal Presentation" sur le memoire de la FTQ, September 12, 1979.

United Steel Workers of America
1290 St. Denis Street
10th floor
Montreal, Quebec H2X 3J7
Telephone: (514) 465-2454

Contact:
> Serge Trudel, Representative

Objectives/Goals:
> Information available upon request.

Research/Projects:
> Information available upon request.

Publications:
> Bulletin Santé et Sécurité.
>
> List available upon request.

Medical

Clinique de Medicine du Travail de Montreal
No. 305, 234 est. boul. Dorchester
Montreal, Quebec H2X 1N8
Telephone: (514) 871-1386

Contact:
: Dr. Michael Lesage, President/General Manager

Objectives/Goals:
: Medical consultant in occupational safety and health care.

Research/Projects:
: Medical and epidemiological surveillance of workers in plants.

    Organizes and manages medical programs.

Publications:
: None.

MEXICO

Academic

Universidad Autonoma Metropolitana-xochimilco
Maestria en Medicina Occupacional
Facultad de Medicina
Depto. de Medicine Soc. Med. Prev. y Salud Publica
Universidad Nacional Autonoma de Mexico
Ciudad Universitaria, D.F., Mexico
Telephone: 594-7833 x. 205

Contact:
: Jorge R. Fernandez Osorio, M.D.

Objectives/Goals:
: Masters-level training in occupational health for health professionals.

Research/Projects:
: Information available upon request.

Publications:
: Information available upon request.

Universidad Autonoma Metropolitana-xochimilco
Calzada del Hueso y Canal Nacional
Mexico 23 D.F.
Telephone: 594-7833, 594-7002

Contact:
> Dr. Miguel Arenas Vargas, Director, Division of Biological Sciences and Health

Objectives/Goals:
> Formation of human resources in teaching, research, and direct services in the fields of occupational health.

Research/Projects:
> Ongoing: Research on working conditions in small and large industries.
>
> Prospective: Industrial fatigue as an occupational illness; interdisciplinary research on health; Master's Degree Program in occupational health.
>
> Past: social service program in industry; physiological, medical, and psychiatric study of workers exposed to electrical hazards.

Publications:
> Information available upon request.

University of Mexico
Centro de Ciencias de la Atmosfera Cd.
Universitaria
Mexico 20 D.F.
Telephone: 5-48-81-90

Contact:
> Armando P. Baez, Researcher

Objectives/Goals:
> Research in water pollution, wastewater treatment, and reverse impact of pollution on biological ecosystems.

Research/Projects:
> Evaluation of effects in the biota at Chachalacas Lagoon due to industrial liquid discharges into the lagoon from a sugarcane refining factory.
>
> Mercury residues contamination in estuaries.
>
> Current research in the chemistry of rain water precipitation in Mexico.

Publications:
> Environmental Impact Due to Industrial Chromium Residues at Lecheria Town, State of Mexico, Mexico.
>
> Movements and Fate of Mercury in an Aquatic Ecosystem.
>
> Aquatic Organism Contamination by Mercury Residues in the Coatzocoalcos River Estuary, Mexico.
>
> Changes in Water Quality at Xochimilco Lake Due to Raw Municipal Sewage Discharge into the Lake.

UNITED STATES

Arizona

Academic

University of Arizona Health Sciences Center
The Arizona Center for Occupational Safety and Health
Tucson, Arizona 85724

Contact:
> H. Abrams, M.D.

Objectives/Goals:
> Education, training, and research in occupational safety and health topics.

Research/Projects:
> Information available upon request.

Publications:
> Information available upon request.

California

Academic

University of California at Berkeley
Labor Occupational Health Program
Institute of Labor Relations
Berkeley, California 94720

Contact:
> Morris Davis, Director

Objectives/Goals:
> To provide education and training programs for workers on occupational safety and health problems.

Research/Projects:
> Provides consultations, conferences, workshops, and long-term training sessions for industries, including: foundries, construction, chemicals and allied product manufacturers.

Publications:
> A Worker's Guide to Documenting Health and Safety Problems.
>
> Toxic Substances Regulated by OSHA: A Guide to Their Properties and Hazards.
>
> Working for Your Life (film), Andrea Hrikko and Melanie Brunt.

Government

California OSHA,
Division of Occupational Safety and Health
455 Golden Gate
San Francisco, California 94101
Telephone: (415) 557-2327

Contact:
> Arthur Carter, Chief

Objectives/Goals:
> Regulatory agency on worker health and safety.

Research/Projects:
> Broad range of projects on specific toxic substances for standard setting and epidemiological monitoring.

Publications:
> List available upon request.

Health

San Francisco General Hospital
Occupational Health Clinic ("Workers Clinic")
1001 Potrero Avenue
San Francisco, California

Contact:
    Molly Coye, Vivian Lin, Linda Morse

Objectives/Goals:
    To serve the needs of workers who experience occupational health problems or who believe that they may have such problems.

    To provide medical, industrial hygiene, legal, and educational services for labor unions and unorganized workers. Clinic meets every Tuesday night.

Research/Projects:
    Labor outreach work, staff education on various occupational health problems.

Publications:
    Information available upon request.

Labor/Public Interest

UNION W.A.G.E. (Union Women's Alliance to Gain Equality)
Health and Safety Committee
37-A 29th Street
San Francisco, California 94010
Telephone: (415) 282-6777 (leave message)

Contact:
    Ellen Bernstein or Karen Garrison

Objectives/Goals:
    To gather and share information about occupational hazards to women workers—focus on clerical workers' hazards, especially stress-related hazards.

    To build an organization of working women to challenge these hazards.

Research/Projects:
> Question and Answer Column in bimonthly UNION W.A.G.E. paper.
>
> Health and Safety Survey of office workers in the San Francisco Bay area.
>
> Educational workshops on hazards for unions and independent groups of clerical workers.

Publications:
> Union W.A.G.E.

Public Interest

Friends of the Earth (FOE)
124 Spear Street
San Francisco, California 94105
Telephone: (415) 495-4770

Contact:
> Jeffrey Knight

Objectives/Goals:
> To promote restoration and preservation of the ecosphere. Issues of concern include: energy sources, alternative energy, conservation, soft energy paths, public lands, wildlife and fisheries preservation and management, workplace and agricultural health, pesticides, recombinant DNA, citizen involvement in government.

Publications:
> Not Man Apart, general environmental news, monthly to members ($25/year).
>
> Soft Energy Notes, bimonthly ($25) published by FOE's international Project for Soft Energy Paths, current information on soft energy developments.
>
> List available upon request.

Indian Training Network
36292 Berkshire Place

Newark, California 94560
Telephone: (415) 797-8341

Contact:
    Joan Bordman, Coordinator

Objectives/Goals:
    To provide culturally relevant training programs and social services for American Indian people.

    To identify specific training needs pertinent to local groups, communities, and individuals with whom Indians interface.

Research/Projects:
    To identify culturally relevant training models for the development of programs and staff responsible for environmental safety and health issues.

    To identify and employ training instructors.

Publications:
    None.

Institute for Food and Development Policy
2588 Mission Street
San Francisco, California 94110
Telephone: (415) 648-6090

Contact:
    Frances Moore Lappe

Goals/Objectives: Nonprofit research, documentation, and education center.

    Cuts through popular myths about hunger, using books, articles, media, and speakers.

    Helps people find appropriate responses to world hunger by assessing U.S. foreign assistance programs and offering guidance on current antihunger work in the United States.

    Helps people weigh alternative long-term strategies to end hunger.

Helps build networks of progressive groups worldwide by making contact with people in all parts of the world who are working to end hunger in their countries.

Research/Projects:

The Aid Debate: to gather evidence on the impact of U.S. aid programs and provide a framework to evaluate the role of aid programs against other U.S. policies.

What Can We Do?: to examine the strengths and weaknesses of various food-related research projects in the United States.

Food Security Project: to examine the ways in which popular decision making can interface with regional and national planning to assist groups working for food security.

Myths of U.S. Food Security: to work with other food policy groups on a critical examination of the U.S. food system.

Publications:

World Hunger: Ten Myths, Frances Moore Lappe and Joseph Collins (72 pages, $2.25).

Needless Hunger: Voices from a Bangladesh Village, Betsy Hartmann and James Boyce (72 pages, $3.00).

Food First Resource Guide, Staff, Institute for Food and Development Policy (80 pages, $3.00).

Agrarian Reform and Counter-Reform in Chile, Joseph Collins (24 pages, $1.00).

Aid to Bangladesh: For Better or Worse?, Michael Scott (28 pages, $1.50).

Tanzania/Mozambique: Lessons, Not Models, Winter 1980.

Myths of Aid, Winter 1980.

What Can We Do?, a guide to action and self-education on food and land, Winter 1980.

82 / SAFETY AND HEALTH CATALOGUE

Public Interest/Media

Occupational Health News
1070 Ardmore Avenue
Oakland, California 94610
Telephone: (415) 465-7257

Contact:
> Vivian Lin

Objectives/Goals:
> Clipping service: to keep members up to date on events and activities in occupational health.

Publications:
> Occupational Health News, $9/6 mos. for clippers, $15/6 mos. for nonclippers.

Public Interest

Public Media Center/Community Occupational Health Project
2751 Hyde Street
San Francisco, California 94109
Telephone: (415) 885-0200

Contact:
> Mary Shinoff

Objectives/Goals:
> To educate unorganized/organized workers on occupational safety and health issues through media workshops, publications, and technical assistance. Current target groups include garment workers, culinary workers, and printers.

Research/Projects:
> Worker attitudes toward occupational health.
>
> Literature review of occupational health problems in garment, culinary, and printing industries.

Publications:
> Looking Out for #1: A Guide to Health Hazards on the Job.
>
> Workers' Rights Handbook.

The Hazard Connection: A Guide to California Resources in Occupational and Environmental Health.

Santa Clara Center for Occupational Safety and Health (ECOSH/PHASE)
655 Castro Street, No. 3
Mountainview, California 94041
Telephone: (415) 969-7233

Contact:
    Robin Baker

Objectives/Goals:
    SantaCOSH is an umbrella organization sponsoring health and safety projects for workers in the Santa Clara Valley of California with emphasis on the electronics industry.

    ECOSH (Electronics Committee on Occupational Safety and Health): a membership organization consisting of workers and technical resource people, developing activist approaches to changing working conditions in the dangerous electronics industry.

    PHASE (Project on Health and Safety in Electronics): provides hazards research, technical assistance, worker training, and other technical services.

Publications:
    ECOSH News, monthly.

    PHASE Fact Sheets (toxic substances in the electronics industry, TCE hazards, reproductive hazards in electronics, fluxes, and solders, and so on).

    PHASE-Profile of the Electronics Workforce in Silicon Valley.

Sierra Club
U.S. Office
530 Bush Street
San Francisco, California 94108
Telephone (415) 981-8634

International Office
800 Second Avenue

84 / SAFETY AND HEALTH CATALOGUE

New York, New York 10017
Telephone: (212) 867-0080

Contact:
    (in New York) Program Associate; (in San Francisco)
    Carl Pope, Labor Liaison, Conservation Department

Objectives/Goals:
    To explore, enjoy, and protect the world's natural heritage.

Research/Projects:
    Industrial safety standards relating to issues of environmental health.

    Public education and lobbying activities.

    Symposia in cooperation with European organizations on the harmonization of National Toxics Regulations.

    Nuclear waste management task force.

Publications:
    Fact sheets available on various aspects of toxics regulation.

    List available upon request.

Southeast Asia Resource Center
P. O. Box 4000D
Berkeley, California
Telephone: (415) 548-2546

Contact:
    Rachel Grossman

Objectives/Goals:
    Nonprofit resource and education center covering events in Southeast Asia with a focus on U.S. involvement in that region.

Research/Projects:
    Research paper on the impact of the U.S. electronics industry in Asia.

    Contacts with activists and researchers concerned with health and safety issues.

Publications:
> Southeast Asia Chronicle No. 66, <u>Changing Role of Southeast Asian Women</u> (focuses on U.S.-owned electronics industry, covering many issues including health and safety hazards).

System Safety Society
29504 Whitley Collins
Rancho Palos Verdes, California 90274
Telephone: (213) 541-1570

Contact:
> David E. Galas, Technical Editor, <u>Hazard Prevention</u>

Objectives/Goals:
> To advance the state-of-the-art of System Safety.
>
> To contribute to a meaningful management and technological understanding of System Safety.
>
> To disseminate newly developed knowledge to all interested groups and individuals.
>
> To further the development of the professionals engaged in System Safety.
>
> To improve public understanding of the System Safety discipline.
>
> To improve the communication of the System Safety movement and discipline to all levels of management, engineering, and other professional groups.

Research/Projects:
> International System Safety Conferences are sponsored biennially. These conferences have proven to be a very popular and effective means for highlighting the latest techniques, applications, and social-legal aspects of System Safety. Minisymposia are sponsored by local chapters, providing an in-depth exploration of a specific System Safety-related topic.
>
> Chapter dinner meetings, field trips, and panel discussions are held at intervals throughout the year.

The Society is cosponsor of various System Safety-related symposia and conferences.

Publications:

Hazard Prevention is the official Society journal. Published five times a year, it keeps members informed of the latest developments in the field of System Safety.

Chapter Newsletters are published periodically to disseminate news of chapter activities and items of interest to chapter members.

Proceedings of Society-sponsored conferences and symposia are made available to members at special discount.

Who Can Join:

All persons engaged in a professional practice related to the art, science, or technology of System Safety, in work that contributes to the advancement of the System Safety concept, or enrolled in courses covering System Safety principles are eligible for membership in the Society.

Western Institute for Occupational and Environmental Sciences (WIOES)
2520 Milvia
Berkeley, California 94704
Telephone: (415) 845-6479

Contact:
Phillip Polakoff, Director

Objectives/Goals:
Clinical research, education, and services on occupational and environmental issues.

Research/Projects:
Clinical and psychological effects of asbestos exposure.

Air traffic controllers.

Worker notification programs.

Publications:
Slide show "Are you Dying for a Job . . . ?"

Pamphlets on Asbestos.

Audiocassettes on Occupational Health.

Connecticut

Academic

University of Connecticut Health Center,
New Directions Program
Farmington, Connecticut 06052
Telephone: (203) 674-2458

Contact:
    Dr. Ray Elling, Director

Objectives/Goals:
    To educate and train workers in the state of Connecticut about occupational safety and health concerns.

Research/Projects:
    Provides free courses throughout the state of Connecticut on occupational safety and health in the following areas: OSHA laws; physiology; noise; stress; ventilation; women's occupational health; health and safety committees; contract language for bargaining agreement; training session with shop stewards.

Publications:
    New Directions News, quarterly publication.

    List of slide shows, films, slide/tapes, and related research material available upon request.

Yale University School of Medicine,
Department of Epidemiology and Public Health
60 College Street
New Haven, Connecticut 06510
Telephone: (203) 436-2422

Contact:
    George A. Silver, M.D., Professor

88 / SAFETY AND HEALTH CATALOGUE

Objectives/Goals:
  Teaching, research, and policy studies.

Research/Projects:
  Child health program analysis.

Publications:
  Child Health: America's Future.

  A Spy in the House of Medicine.

District of Columbia

Government

Consumer Product Safety Commission (CPSC)
111 18th Street, N.W.
Washington, D.C. 20207
Telephone: (800) 638-8326

Contact:
  John Bell, Office of Media Relations

Objectives/Goals:
  PSC is an independent federal regulatory agency mandated to reduce the number of accidents and deaths that result from consumer products.

Research/Projects:
  Enacts mandatory safety standards.

  Participates in industry efforts to develop voluntary safety standards.

  Bans products for which no feasible safety standard would adequately protect the public.

  Seeks, negotiates, and monitors corrective action plans for products that may present a substantial hazard to consumers.

  Informs and educates consumers about product hazards.

  Conducts research and develops test methods.

Collects and publishes injury and hazard data.

Promotes uniform product regulations by government units.

Publications:
Information available upon request.

The Council on Environmental Quality (CEQ)
722 Jackson Place, N.W.
Washington, D.C. 20006
Telephone: (202) 633-7027

Objectives/Goals:
Established in the executive office of the president with its primary responsibility to provide Congress with an annual Environmental Quality Report emphasizing the nation's environment.

Research/Projects:
To analyze and coordinate federal environmental policy and advise the president on environmental matters. Further information available upon request.

Publications:
Information available upon request.

Environmental Protection Agency (EPA)
401 M Street, S.W.
Washington, D.C. 20460
Telephone: (202) 755-0700

Objectives/Goals:
To protect the nation's land, air, and water systems. Under a mandate of national environmental laws focused on air and water quality, solid waste management, and the control of toxic substances, pesticides, noise and radiation, the agency strives to formulate and implement actions that lead to a compatible balance between human activities and the ability of natural systems to support and nurture life.

Research/Projects:
Research in all of the above areas.

Publications:
Contact EPA.

Occupational Safety and Health Administration (OSHA)
200 Constitution Avenue, N.W.
Washington, D.C. 20210
Telephone: (202) 523-8013

Objectives/Goals:
OSHA is the federal agency within the Department of Labor that sets and enforces job safety and health standards in the private sector. In addition, the agency promotes voluntary compliance with its workplace regulations through a variety of consultation, training, education, and information programs for workers and management. OSHA approves and monitors the operations of states that have chosen to administer their own occupational safety and health programs. The agency was created by the Occupational Safety and Health Act of 1970 and currently has jurisdiction over about 5 million workplaces where approximately 62 million workers are employed.

Research/Projects:
Information available upon request.

Publications:
OSHA offers a number of free pamphlets describing the agency, as well as various technical and scientific booklets on occupational safety and health. A Publications Catalogue may be ordered by writing: OSHA Publications, Room S-1212, Washington, D.C. 20210. Further information can be found at the following regional offices for OSHA:

Region 1 (Connecticut, Maine, Massachusetts, New Hampshire, Rhode Island, Vermont)
JFK Federal Bldg., Room 1804
Government Center
Boston, MA 02203

Region II (New Jersey, New York, Puerto Rico, Virgin Islands, Canal Zone)

Room 3445, 1 Astor Place
1515 Broadway
New York, NY 10036

Region III (Delaware, District of Columbia, Maryland,
Pennsylvania, Virginia, West Virginia)
Gateway Bldg., Suite 2100
3535 Market St.
Philadelphia, PA 19104

Region IV (Alabama, Florida, Georgia, Kentucky,
Mississippi, North Carolina, South Carolina, Tennessee)
1375 Peachtree St., N.E.
Suite 587
Atlanta, GA 30309

Region V (Illinois, Indiana, Michigan, Minnesota, Ohio,
Wisconsin)
230 S. Dearborn St.
32nd Floor, Room 3263
Chicago, IL 60604

Revion VI (Arkansas, Louisiana, New Mexico, Oklahoma,
Texas)
555 Griffin Square, Room 602
Dallas, TX 75202

Region VII (Iowa, Kansas, Missouri, Nebraska)
911 Walnut St., Room 3000
Kansas City, MO 64106

Region VIII (Colorado, Montana, North Dakota, South
Dakota, Utah, Wyoming)
Federal Bldg., Room 1554
1961 Stout St.
Denver, CO 80294

Region IX (Arizona, California, Hawaii, Nevada)
Box 36017
450 Golden Gate Ave.
San Francisco, CA 94102

Region X (Alaska, Idaho, Oregon, Washington)
Federal Office Bldg., Room 6002
909 First Ave.
Seattle, WA 98174

92 / SAFETY AND HEALTH CATALOGUE

Health

National Council for International Health
2121 Virginia Avenue, N.W.
Suite 303
Washington, D.C. 20037
Telephone: (202) 338-1142/3

Contact:
> Sherri A. Simches, Information and Communications Specialist

Objectives/Goals:
> Information available upon request.

Research/Projects:
> Information available upon request.

Publications:
> List available upon request.

Society for Occupational and Environmental Health
1341 G. Street, N.W.
Suite 308
Washington, D.C. 20005
Telephone: (202) 347-4550

Objectives/Goals:
> To promote public attention on the social, scientific, and regulatory problems involved in improving the quality of both working and living places.

Research/Projects:
> Sponsors conferences, workshops, and publications.

Publications:
> List available upon request.

Labor

African-American Labor Center
1125 15th Street, N.W.
Suite 404

Washington, D.C. 20005
Telephone: (202) 293-3603

Contact:
Lester Trachtman

Objectives/Goals:
To assist the African trade unions to better serve their members and contribute to the development of their countries.

Research/Projects:
Sponsored Pan-African Conferences on Occupational Health and Safety of Agricultural Workers (Addis Ababa, 1973) and Occupational Health and Safety of Mine Workers (Zambia, 1976).

Provided occupational health training for African mine workers in the United States.

Publications:
"Mine" Conference Reports.

Industrial Union Department, AFL-CIO
815 16th Street, N.W.
Washington, D.C. 20006
Telephone: (202) 393-5581

Contact:
Sheldon W. Samuels, Director, Health, Safety, and Environment

Objectives/Goals:
A central labor group serving 6 million workers in 58 unions.

Research/Projects:
Information available upon request.

Publications:
Information available upon request.

Labor/Public Interest

National Association of Farmworker Organizations (NAFO)
1332 New York Avenue, N.W.
Washington, D.C. 20005
Telephone: (202) 347-3407

Contact:
> Maria Mazorra, Industrial Hygienist

Objectives/Goals:
> Nonprofit organization to protect the interests and legal rights of farmworkers (unions, migrant clinics, community and legal groups).

Research/Projects:
> Information available upon request.

Publications:
> Pesticide Manual.
>
> Pesticide Educational Slide Show.
>
> VIDEO presentations on victims of pesticide poisonings.

Labor

United Auto Workers
International Affairs Department
1757 N Street, N.W.
Washington, D.C. 20036

Contact:
> John Christensen

Objectives/Goals:
> International Trade Union Occupational Health and Safety Programs.

Research/Projects:
> (1980) Occupational Safety and Health Seminars in Colombia, Brazil, Spain, Portugal.

Publications:
> <u>UAW Worldwide Newsletter.</u>

<u>UAW Washington Report</u>.

<u>UAW Solidarity</u>.

Public Interest

American Labor Education Center
1835 Kilbourne Place, N.W.
Washington, D.C. 20010
Telephone: (202) 465-8925

Contact:
    Matt Witt

Objectives/Goals:
    Nonprofit institution providing educational programs and materials on labor-related problems to workers, unions, and the general public.

Research/Projects:
    Research for monthly newsletter and education guide, <u>American Labor</u>.

    Researching the wood products industry in Sweden, West Germany, and Austria with members of the International Woodworkers of America.

Publications:
    <u>American Labor</u>, monthly newsletter.

American Public Health Association (APHA)
1015 Fifteenth Street, N.W.
Washington, D.C. 20005
Telephone: (202) 789-5600

Contact:
    Larry J. Gordon, M.S., M.P.H., President

Objectives/Goals:
    APHA is devoted to the protection and promotion of public health.

96 / SAFETY AND HEALTH CATALOGUE

Research/Projects:
> Initiates projects designed for improving health both nationally and internationally.
>
> Researches health problems and offers possible solutions based on that research.
>
> Launches public awareness campaigns about specific health dangers.
>
> Sets standards for alleviating health problems.
>
> Publishes numerous materials reflecting the latest findings and developments in public health.
>
> Carries a year-round schedule of action implementation programs.

Publications:
> American Journal of Public Health (monthly newspaper on current health legislation and policy issues).
>
> The Nation's Health (monthly newspaper on current health legislation and policy issues).
>
> Washington Newsletter (summary of all health-related legislative activities from federal agencies).
>
> APHA Publications: List available upon request.

Center for Science in The Public Interest
1757 S. Street N.W.
Washington, D.C. 20009
Telephone: (202) 332-4250

Objectives/Goals:
> Nonprofit organization to provide research and resources on national and local levels in the areas of energy and nutrition.

Research/Projects:
> Information available upon request.

Publications:
> People and Energy, monthly newsletter.

Nutrition Action, monthly newsletter.

Additional list available upon request.

The Coalition for the Reproductive Rights of Workers (CRROW)
1126 16th Street, N.W. no. 316
Washington, D.C. 20036
Telephone: (202) 659-1311

Objectives/Goals:
    CRROW was formed in 1979 to defend the employment rights and reproductive freedom of workers who are exposed to toxic chemicals and other hazards. It is composed of labor unions, legal organizations, women's groups, and environmental groups.

    CRROW seeks an end to the unacceptable choice between a job and the right to reproduce. It is working to eliminate the corporate policy of altering the worker rather than the workplace, to inform workers of their rights, and to lead the fight for clean and healthy workplaces that insure workers' reproductive freedom.

Research/Projects:
    Information available upon request.

Publications:
    List available upon request.

Commission for the Advancement of Public Interest Organizations
1875 Connecticut Avenue, N.W.
Suite 1013
Washington, D.C. 20009
Telephone: (202) 462-0505

Contact:
    Suzanne Harmon

Objectives/Goals:
    Established by the Monsonn Medical Foundation to further the aims of the public interest movement.

Research/Projects:
> Workshops and conferences on various public interest issues.
>
> Provides information and written statements to legislators and government agencies on a variety of public interest concerns.
>
> Provides computer information system on organizations, resources, publication references, and government documents on a variety of national and international concerns.

Publications:
> Information available upon request.

The Conservation Foundation
1717 Massachusetts Avenue, N.W.
Washington, D.C. 20036
Telephone: (202) 797-4300

Objectives/Goals:
> A nonprofit research and communication organization dedicated to encouraging human conduct to sustain and enrich life on earth. It has attempted to provide intellectual leadership in the cause of wise management of the earth's resources.

Research/Projects:
> Land use and urban growth.
>
> Coastal resources management.
>
> Public lands.
>
> Citizen training, water quality, energy conservation.
>
> Environmental ethics and economics.

Publications:
> CF Newsletter, monthly newsletter.
>
> List of publications and films available upon request.

Environmental Action Foundation
724 Dupont Circle Building
Washington, D.C. 20036
Telephone: (202) 659-9682

Contact:
　　Helen Sandalls, Toxics Project Coordinator

Objectives/Goals:
　　To provide concerned citizens with information on the problems associated with toxic chemicals and the means to fight involuntary exposure to toxic chemicals.

Research/Projects:
　　To provide a clearinghouse on environmental topics.

Publications:
　　Newsletter, focusing on toxic contamination.

　　Fact sheets.

　　Resource guide.

　　Experts list (referrals for grassroots groups).

The Environmental Defense Fund (EDF)
1525 18th Street, N.W.
Washington, D.C. 20036
Telephone: (202) 833-1484

Contact:
　　Joseph H. Highland, Ph.D.

Objectives/Goals:
　　To preserve and enforce important environmental laws.

Research/Projects:
　　Combining scientific, economic, and legal expertise in the following programs: wildlife program; energy program; toxic chemicals program; water resources program.

Publications:
　　EDF Newsletter.

Malignant Neglect, by EDF and Robert H. Boyle (New York: Alfred A. Knopf, 1979).

List available upon request.

Environmentalists for Full Employment
1101 Vermont Avenue, N.W.
Suite 305
Washington, D.C. 20005
Telephone: (202) 347-5590

Contact:
Richard Grossman

Objectives/Goals:
National environmental organization to promote the concept that there need be no conflict between our nation's providing safe, socially useful jobs and healthy community and natural environments.

Research/Projects:
Information available upon request.

Publications:
Jobs and Energy, 1977. Examination of the employment impacts of various energy technologies.

List available upon request.

Environmental Law Institute
1346 Connecticut Avenue, N.W.
Suite 600
Washington, D.C. 20036
Telephone: (202) 492-9600, extension 237

Contact:
Devra Davis, Program Director

Objectives/Goals:
Research and analysis of environmental topics, including environmental and occupational health.

Research/Projects:
Current research programs include: energy; toxic substances; air/water; resources; land use.

Publications:
List available from Environmental Law Institute.

Environmental Policy Center
317 Pennsylvania Avenue, S.E.
Washington, D.C. 20003
Telephone: (202) 547-6500

Contact:
Louise Dunlap, Executive Vice-President

Objectives/Goals:
The center maintains a large team of public interest lobbyists working on energy and water resource policies. Its major objectives are to promote energy conservation, renewable energy resources, diverse/decentralized/regionalized energy production systems; while working to reform nuclear siting and licensing procedures, federal radiation standards and transportation/decommissioning/waste disposal policies within the fossil fuel sector (coal/oil/gas), the center is working to minimize adverse health environmental and social impacts of specific types of energy development.

Research/Projects:
Nuclear: Low-level radiation, uranium enrichment, nonproliferation, export-import bank policies, SALT.

Nuclear: waste disposal, transportation, decommissioning.

Water: president's national water policy; transportation (navigation/rails); corps of engineers/Bureau of Reclamation projects; water user conflicts.

Energy Conservation and Alternative Technologies: Department of Energy budget; solar power satellite; conservation (industrial/transportation/residential); energy trust fund.

Rural Energy/Agricultural Resources: Rural electrification administration/federal power administrations; pipeline/

powerline siting; 1872 Mining Act reform; agricultural impacts of energy development.

Coal: Strip-mining/federal leasing congressional oversight. Nuclear: siting and licensing reform; demand forecasting; Price Anderson reform; decentralized facility siting alternatives.

Oil and gas: Oil export/import policies; oil spill liability; liquified petroleum gas siting and safety; oil pipeline siting and licensing; outer continental shelf oil and gas congressional oversight.

Nonfuels Mineral Resources: Deep seabed mining.

Publications:
List available upon request.

Environmental Policy Institute
317 Pennsylvania Avenue, S.E.
Washington, D.C. 20003
Telephone: (202) 544-8200

Contact:
Louise Dunlap, President

Objectives/Goals:
A research, educational, and executive branch policy development organization focusing on national energy and water resource public policies.

Research/Projects:
Citizens Coal Project.

Federal Coal Leasing.

Citizens Oil and Gas Project.

Energy Conservation Project.

Energy Facility Siting Analysis Project.

Radiation Health Information Project.

Nuclear Waste Disposal and Transportation.

Rural Land and Energy Project.

Water Conservation Project.

Publications:
List available upon request.

Environmental Policy Institute
Radiation Health Information Project
317 Pennsylvania Avenue, S.E.
Washington, D.C. 20003
Telephone: (202) 544-8200

Contact:
Robert Alvarez, Elli Walters, Director/Assistant Director

Objectives/Goals:
A public interest research and information organization specializing in energy and natural resources public policy issues.

The center is working to promote the following: energy conservation; renewable energy resources, diverse/decentralized/regionalized energy production systems; reform nuclear siting and licensing procedures and transportation/decommissioning/waste disposal policies.

Research/Projects: Current research includes:
Radiation Health Information Project: to stimulate research on health hazards from radiation; to expand public awareness and understanding of radiation; to provide policy alternatives for federal/state radiation programs.

Occupational and public exposure standards.

Scientific and public debate over radiation.

Effects of radiation on our national energy policy.

Kerr McGee former nuclear worker health follow-up.

Analysis of Hanford Nuclear Workers data.

Nonnuclear fuel cycle occupational radiation exposures.

Publications:
List available upon request.

Equity Policy Center
1302 18th Street, N.W.
Suite 502
Washington, D.C. 20036
Telephone: (202) 223-8170; Cable: EPOC Washington, D.C.

Contact:
Irene Tinker, Director

Objectives/Goals:
A nonprofit research, communications, and educational group to study and promote means toward more equitable distribution of income and resources at home and abroad.

Research/Projects:
Presently organizing an international symposium on women and their health.

Policy paper on food technology and women.

Exploratory symposium and roundtables on women and power.

Workshop on women and development, held by AAAS as a contribution to the U.S. preparation for the UN Conference on Science and Technology for Development.

Policy paper on women and development issues.

Socioeconomic considerations for energy programs.

Publications:
"Socio-Economic Considerations for Household and Rural Energy Programs: Notes on Field Trips to India and the Philippines," Dr. Irene Tinker.

"New Technologies for Food-Related Activities: An Equity Strategy," Dr. Irene Tinker.

"Technology, Poverty and Women: Some Special Issues," Dr. Irene Tinker.

"Changing Energy Usage for Household and Subsistence Activities," Dr. Irene Tinker.

International Commission for Occupational and Environmental Health
1341 G Street, N.W.
Washington, D.C. 20005
Telephone: (202) 347-4550

Contact:
   Knut Ringen, Executive Secretary

Objectives/Goals:
   To promote reduction of health effects associated with international technology transfer through improvements of national occupational health standards and regulations.

Research/Projects:
   Ongoing program of expert meetings and conferences.

Publications:
   Proceedings of Workshop on National Occupational Health Standards, Summer 1980.

International Institute for Environment and Development (IIED)
1302 Eighteenth Street, N.W.
Suite 501
Washington, D.C. 20036
Telephone (202) 462-0900

27 Mortimer Street
London W1N 8 DE
United Kingdom
Telephone: 01-580-7656-7

Contact Barbara Ward, President

Objectives/Goals:
   Nonprofit organization that serves as a catalyst for information action on environmental and developmental problems in the international community.

Research/Projects:
> Dissemination of ideas to governments and international agencies to assist their formulation of policies at regional and global levels.
>
> Organization of expert consultant groups that focus attention on specific problems and suggest their solutions as well as methods to achieve them.
>
> Arousal of both public and nongovernmental organizations' interest in the policies needed to solve particular world problems and assisting such groups to organize their efforts around these issues.
>
> Earthscan, an independent environmental information service whose task is to make the world's media aware of important issues centered around environment and development.

Publications:
> List available upon request.

National Wildlife Federation (NWF)
1412 16th Street, N.W.
Washington, D.C. 20036
Telephone: (202) 797-6883

Contact:
> Sandra Jerabek, Waste Project Director; Anne Sidbury; Kathy Painter; Ken Kamlet

Objectives/Goals:
> American conservation organization dedicated to the wise use and conservation of our natural resources. NWF has affiliate groups in every state and a variety of programs, including areas such as toxics. In particular, the NWF's Waste Alert! program is conducting information and outreach to public interest groups across the country through a series of regional and state workshops. This program is building a coalition with diverse groups, including labor, on toxics and hazardous waste issues.

Publications:
> Toxic Substances Programs in the U.S. States and Terri-

tories: How Well Do They Work? (single copies available free of charge).

State Background Papers Assessing Hazardous Waste Management; NWF Waste Project; available for selective regions of the country.

Natural Resources Defense Council (NRDC)
917 15th Street, N.W.
Washington, D.C. 20005
Telephone: (202) 737-5000

Contact:
    Jacob Scherr, Esq.

122 E. 42nd Street
New York, NY 10017
Telephone: (212) 949-0049

Contact:
    Carol Hine

Objectives/Goals:
    Nonprofit national legal group that litigates on environmental issues, participates in administrative proceedings, and directs public educational projects.

Research/Projects:
    Current projects include: toxic substances, energy air and water pollution, coastal zone preservation, transportation, international toxics, short wave radiation.

Publications:
    NRDC Newsletter.

    AMICUS (monthly publication).

Public Citizen's Health Resource Group
2000 P. Street, N.W.
Washington, D.C. 20036
Telephone: (202) 872-0320

Contact:
    Sidney M. Wolfe, M.D., Director

108 / SAFETY AND HEALTH CATALOGUE

Objectives/Goals:
> Nonprofit consumer advocacy group funded through private contributions made to Ralph Nader's Public Citizen, Inc.

Research/Projects:
> Research areas include: food and drug regulation; workplace safety and health; product safety; health care delivery systems.

Publications:
> List of publications available upon request.

Technical Information Project
1346 Connecticut Avenue, N.W.
Room 217
Washington, D.C. 20036
Telephone: (202) 466-2954

Objectives/Goals:
> Research on energy, wastes, and toxic substance problems and policies. Sponsors interdisciplinary conferences in these areas.

Research/Projects:
> Conference series with topics (to date): trade secrecy; decision making; information access; compensation and global environmental problems.

Publications:
> Pamphlet on solid waste issues of citizen concern.
>
> Conference proceedings.

Urban Environment Conference (UEC)
1302 18th Street, N.W.
Room 301
Washington, D.C. 20036
Telephone: (202) 466-6040

Contact:
> George Coling, Coordinator

Objectives/Goals:
> To build and service coalitions between labor, minorities, and environmentalists on urban environmental issues of common concern. Chief priority is occupational and environmental health, but research work is done in areas of transportation reform, energy policy, and fair housing. Both citizen/worker education and lobbying programs are used to build and service these organizations.

Research/Projects:
> Occupational Health Resource Center: occupational health and safety education programs for women and minority workers. The program works through local community-based organizations.
>
> Regional Workshops Program sponsors regional conferences nationwide on issues of toxic substances control and environmental and occupational health.
>
> Lobbying efforts for strong occupational and environmental health laws, transportation reform, and fair housing.
>
> Networking among and assistance to local coalitions and affiliated organizations.

Publications:
> Environmental Cancer: Causes, Victims, Solutions ($1.50).
>
> Inner City Health in America (published by the Urban Environment Foundation affiliate, $3.00).
>
> A Worker's Guide to NIOSH (free).
>
> A Guide to Worker Protections in Environmental Laws (free).
>
> A Guide to Worker Education Materials in Occupational Safety and Health (free).

Workers' Institute for Safety and Health
1126 16th Street, N.W.
Washington, D.C. 20036
Telephone: (202) 887-1980

Contact:
> Joseph Valesquez, Executive Director

Objectives/Goals:
>Nonprofit corporation providing research for the American Labor Movement in the areas of health, safety, and the environment. The Institute aims to identify the unique health needs of workers and their families and the available community resources to alleviate those needs.

Research/Projects:
>Develop a technical information service for the labor unions' health programs.
>
>Identify resources for advanced training in health care for labor union representatives.
>
>Community organization to promote workers' rights in health and safety.
>
>Mass media programs to project occupational health and safety concerns.
>
>Follow-up programs to cohorts of workers who are at high risk of disease due to past exposure.
>
>Various categorical research programs related to special hazards.

Publications:
>List available upon request.

Illinois

Academic

University of Illinois
Chicago Labor Education Program
Room 1315 SEO Building
P.O. Box 4348
Chicago, Illinois 60680
Telephone: (312) 996-2623

Contact:
>Helen Elkiss

University of Illinois
Institute of Labor and Industrial Relations (Downstate)

504 East Armory Avenue
Champaign, Illinois 61820

Contact:
    Ronald J. Peters

Objectives/Goals:
    To provide free health and safety training for workers.

Research/Projects:
    Safety and health classes.

    Conferences.

    Seminars and workshops.

Publications:
    Information available upon request.

National Safety Council (NSC)
425 North Michigan Avenue
Chicago, Illinois 60611
Telephone (312) 527-4800

Objectives/Goals:
    Nonprofit, public service organization to provide research and dissemination of information concerning occupational risk prevention.

Research/Projects:
    Risk prevention and accident prevention.

Publications:
    National Safety News, monthly.

    Journal of Safety Research.

    Additional list and further information available upon request.

Public Interest

Citizens for a Better Environment (CBE)
59 East Van Buren Street

112 / SAFETY AND HEALTH CATALOGUE

Chicago, Illinois 60605
Telephone: (312) 939-1984

Objectives/Goals:
> Nonprofit environmental organization providing legal, advisory, and research services for national and local governmental agencies.

Research/Projects:
> Information available upon request.

Publications:
> CBE Environmental Review, monthly.

Indiana

Academic

Indiana University
School of Continuing Studies, Division of Labor Studies
312 North Park
Bloomington, Indiana 47405
Telephone: (812) 337-9082

Contact:
> D. W. Murphy, Director

Objectives/Goals:
> To promote the education and training of students, union leaders, and workers throughout Indiana in labor studies.

Research/Projects:
> Information available upon request.

Publications:
> List available upon request.

Maine

Labor

University of Maine
Bureau of Labor Education

129 College Avenue
Orono, Maine 04473
Telephone: (207) 581-7032

Objectives/Goals:
 To provide educational opportunities and programs of special interest to Maine workers and their organizations.

Research/Projects:
 Educational courses include: Steward training; collective bargaining techniques; labor law; grievance procedure; parliamentary procedure; labor history; public relations in public and private sectors.

 Special program and seminar topics include: Unemployment and Workmens' Compensation; labor economics; consumer affairs; government and the legislative process; the impact of environmental concerns on industrial growth; absentee ownership and its effect on the quality of the workplace; the emergence of Third World countries and world trade.

Publications:
 List available upon request.

Maryland

Government

National Cancer Institute
National Institute of Health
Bethesda, Maryland 20205
Telephone: (800) 638-6694

Contact:
 Robert J. Avery, Jr., Chief, Public Inquiries
 Office of Cancer Communication

Objectives/Goals:
 The principal federal government agency for research on cancer prevention, diagnosis, treatment, rehabilitation, and dissemination of information for the control of cancer.

Research/Projects:
 Cancer research is conducted in the Institute's Bethesda

headquarters and in about 1,000 laboratories and medical centers throughout the United States.

Speakers Bureau provides speakers on all relevant topics, such as environmental carcinogenesis and occupational cancer.

Publications:
List available upon request.

Public Interest

Volunteers in Technical Assistance (VITA)
3706 Rhode Island Avenue
Mt. Rainier, Maryland 20822
Telephone: (301) 277-7000

Objectives/Goals:
Private, nonprofit development organization that supplies information and assistance, primarily by mail, to people and organizations seeking help with technical problems in more than 100 developing countries.

Research/Projects:
Provides worldwide network of experts who have volunteered to respond to requests for assistance for improving homes, farms, communities, businesses, and lives.

Provides assistance to low-income people in their efforts in the areas of agriculture and food, renewable energy sources, shelter, water supply, and small industries.

Designs and adapts tools, methods, programs to respond to local needs, resources, and conditions.

Publications:
Vita News, quarterly newsletter.

Produces handbooks in the areas of wind energy and low-cost construction techniques.

List available upon request.

NORTH AMERICA / 115

Massachusetts

Academic

Boston University Medical Center
Health Policy Institute
Center for Strengthening Health Delivery Systems in Africa
53 Bay State Road
Boston, Massachusetts 02215

Contact:
    Joel Montague, Deputy Director

Objectives/Goals:
    To manage the Agency for International Development funded regional project to strengthen health delivery systems in Central Africa from the university's office in the Ivory Coast.

    To educate and train the indigenous population in primary health care.

Research/Projects:
    To assist African institutions to develop applied research projects in primary health care. The first project will be implemented in 1980.

    To develop training materials in primary health care.

Publications:
    Training materials in primary health care.

Clark University
Center for Technology, Environment, and Development
Worcester, Massachusetts 01610
Telephone: (617) 793-7283

Contact:
    Dr. Roger E. Kasperson, Director

Objectives/Goals:
    Coordination and enlargement of university research effort.

    Development of common theory.

116 / SAFETY AND HEALTH CATALOGUE

Expansion of dissemination and public impact of research findings.

Research/Projects:
Cogeneration Project/University power plant.

Training program in environmental planning for public officials in seven African countries.

Societal management of technological hazards.

Climactic change and population impacts.

Electricity production planning.

Equity issues in radioactive waste management.

Environmental planning in developing societies.

Fuel wood in Kenya.

Differential standards for workplace and the public.

Industrial decline in New England.

Publications:
<u>Network for Environment and Development</u>, newsletter issued by the International Development Research Group.

List available upon request.

Harvard School of Public Health
Occupational Health Program
665 Huntington Avenue
Boston, Massachusetts 02115
Telephone: (617) 732-1260

Contact:
David Wegman, M.D., Associate Professor of Occupational Health

Objectives/Goals:
The aim of our program is to produce persons trained in occupational safety and health who can recognize and

prevent occupational injuries and diseases. We believe that prevention should be the primary orientation of professionals in this area. Today's physicians and nurses frequently operate without public health perspective, and safety engineers frequently know little about occupational health problems. We wish to accomplish our objective by directing our training effort at the development of public health perspectives, the acquisition of skills and knowledge for prevention and the creation of a sensitivity about the political climate in which the professional must act. We therefore intend to train professionals with the necessary technical and social information to teach, do research, administer, consult, or serve in occupational safety and health programs in universities, governmental agencies, private industry, or organized labor to reduce occupational injuries and disease.

Research/Projects:
Ongoing, Past, or Prospective Research/Projects: Our program has long been interested in the relationship between occupational exposures and occupational disease. This interest has been reflected by a series of research efforts aimed at identifying new hazards and bringing them under control. The following is a partial list of exposures and occupational groups that have been or are being studied: asbestos, arsenic coal dust, chlorine, carbon monoxide, carbon black, hemp dust, coal miners, executives, firefighters, granite workers, leather workers, heat, lead, marble, noise, ozone rubber chemicals, TDI, pipecovers, meat packers, transportation workers, talc miners and millers, smelter workers, tire curemen, sulfur dioxide, welders, stress, smoke, vinyl chloride, oxides of nitrogen, tunnel workers.

Publications:
List (including audio/visuals, periodicals) available upon request.

Massachusetts Institute of Technology
Center for Policy Alternatives (CPA)
77 Massachusetts Avenue
E40-250
Cambridge, Massachusetts 02139
Telephone: (617) 253-1663

Contact:
> Dr. Nicholas A. Ashford, Assistant Director, CPA
> Associate Professor of Technology and Policy

Objectives/Goals:
> To identify and study important emerging social issues in which technology and engineering play a significant role.
>
> To assess the consequences of established institutional policies and develop alternatives available to decision makers.
>
> To provide students and professionals with research and training opportunities in policy formulation and analysis.

Research/Projects:
> Science, technology, and public policy.
>
> Consumer policy.
>
> Manufacturing and industrial productivity.
>
> Information and communications systems.
>
> Technological innovation.
>
> Environmental, health, and safety regulation. To study and evaluate the effects of regulation on workers, citizens, and the manufacturing process.

Publications:
> List available upon request.

Health/Public Interest

Boston Women's Health Book Collective
Box 192
West Somerville, Massachusetts
Telephone: (617) 924-0271

Contact:
> Judy Norsigian, Administrative Coordinator

Objectives/Goals:
> Education, advocacy, and activism in all areas of women's health.

Research/Projects:
> Spanish-language edition of Our Bodies, Ourselves.
>
> International Women and Health Resource Guide with ISIS (Women's International Information and Communication Service, Rome, Italy).
>
> Health education project with the Porcupine Women's Health Collective, a group of native American women in the Wounded Knee area.
>
> Speakers Bureau.
>
> Health education programs with Women's Community Health Center, Cambridge, Mass.; and the Somerville Women's Health Project, Somerville, Mass.
>
> Bimonthly women's health information packets.

Publications:
> Our Bodies, Ourselves (New York: Simon and Schuster, 1976).
>
> Nuestros Cuerpos, Nuestras Vidas, 1977.
>
> Ourselves and Our Children (New York: Random House, 1978).
>
> International Women and Health Resource Guide, June 1980.
>
> Healthright, national women's health quarterly.
>
> "Taking Our Bodies Back," Cambridge Documentary Films (film on the women's rights movement).
>
> Health information booklets with CIDHAL, Cuernavaca, Mexico.

120 / SAFETY AND HEALTH CATALOGUE

Public Interest

Massachusetts Coalition for Occupational Safety and Health
718 Huntington Avenue
Boston, Massachusetts 02115
Telephone: (617) 277-0097

Contact:
 Nancy Lessin (staff)

Objectives/Goals:
 To provide technical, legal, and educational support to unions and workers on occupational safety and health problems.

Research/Projects:
 Women's Committees on Health, Technical, and Legal Problems.

 Workshops and educational training programs.

 Women's Occupational Health Conference: "Women's Work; Women's Health," April 1980.

Publications:
 <u>Injured on the Job: A Handbook for Massachusetts Workers</u> (Handbook on worker's compensation).

 "Survival Kit," bimonthly publication.

 Fact Sheets.

New England Human Rights Network
2161 Massachusetts Avenue
Cambridge, Massachusetts 02140
Telephone: (617) 661-6130

Contact:
 Jane Guise, Director

Objectives/Goals:
 To promote economic, social, cultural, civil, and political human rights by sharing information and providing communication and mutual support for individuals and organizations active in all aspects of human rights.

Research/Projects:
> Promoting the implementation of existing human rights amendments to economic and military foreign assistance legislation.
>
> Actively encouraging the development of legislation guaranteeing human rights in all areas of American life.
>
> Working for the U.S. Senate ratification of the UN International Covenants on economic, social, and cultural rights, and on civil and political rights.
>
> Promoting selected other nonviolent human rights action ideas regionwide and coordinated with nationwide activities.
>
> Establishing a monthly newsletter to communicate information about regional activities and needs.
>
> Continuing use of established phone trees within the region for immediate calls to action.
>
> Developing a centralized and current resource file, speakers bureau, and audiovisual list.

Publications:
> Newsletter and list available upon request.

OXFAM-America
302 Columbus Avenue
Boston, Massachusetts 02116
Telephone: (617) 247-3304

OXFAM
274 Banbury Road
Oxford OX27DZ
United Kingdom
Telephone: 0865.56777; Telex: 83610

Contact:
> David Bull

Objectives/Goals:
> Nonprofit, international development agency that funds self-help programs in Asia, Africa, and Latin America.

Economic and food self-reliance are emphasized in all its programs.

Research/Projects:
Currently OXFAM supports or contributes toward over 1,000 projects designed: for the poorest to have more, particularly in terms of food, better health, and a fair share of the world's increasingly limited resources; for the poorest to have more in terms of confidence, relationships, self-determination, and ability to manage their own futures.

Publications:
Extensive list available upon request.

Science for The People
897 Main Street
Cambridge, Massachusetts 02139
Telephone: (617) 597-0370

Objectives/Goals:
To build a science and technology responsive to people's needs.

Research/Projects:
Workshops and study groups on various topics including nuclear power, nutrition, sociobiology, and genetic engineering.

Publications:
Science for The People, bimonthly publication.

Information available upon request.

Union of Concerned Scientists (UCS)
1208 Massachusetts Avenue
Cambridge, Massachusetts 02138
Telephone: (617) 547-5552

Contact:
Eric Van Loon, Director

Objectives/Goals:
Nonprofit group consisting of scientists and engineers

concerned with environmental issues and questions of nuclear power plant safety.

Research/Projects:
Information available upon request.

Publications:
Information available upon request.

Urban Planning Aid
Health and Safety Project
120 Boylston Street
Boston, Massachusetts 02116
Telephone: (617) 482-6695

Objectives/Goals:
To provide technical assistance and resource materials for low-income working people in Eastern Massachusetts.

Research/Projects:
To provide technical information and skills on occupational safety and health problems through workshops, seminars, meetings, and publications.

Publications:
How to Use OSHA, a worker's action guide to the Occupational Safety and Health Act.

How to Look at Your Workplace.

Como Inspeccionar Su Centro de Trabajo (same as above, Spanish edition).

Several fact sheets on various hazards.

Michigan

Academic

University of Michigan
School of Public Health
109 Observatory
Ann Arbor, Michigan 48109
Telephone: (313) 763-5568

Contact:
    Oscar Gish, Lecturer

Objectives/Goals:
    General interest in health and safety and health care issues in the third world.

Research/Projects:
    Information available upon request.

Publications:
    Information available upon request.

Labor

The United Autoworkers (UAW)
Social Security Department
Health and Safety Staff
8000 East Jefferson
Detroit, Michigan 48214
Telephone: (313) 926-5321

Contact:
    Melvin Glasser, Director

Objectives/Goals:
    Representing workers in the automobile, aerospace, and agricultural implement industries, the UAW health and safety program promotes health hazard evaluations, educational programs, collective bargaining, and legislative work.

Research/Projects:
    Information available upon request.

Publications:
    Health and Safety Newsletter.

    UAW pamphlets on occupational hazards.

    List available upon request.

Public Interest

International Association for the Advancement of Appropriate Technology for Developing Countries (IAAATDC)
603 East Madison Street
The University of Michigan
Ann Arbor, Michigan 48109
Telephone: (313) 764-6410

Contact:
    Ike C. A. Oyeka, President

Objectives/Goals:
    Nonprofit interdisciplinary and international association of scholars and students from developing and developed countries interested in the technology transfer problems in developing countries. IAAATDC promotes the socio-economic advancement of developing countries through a systematic application of science and technology appropriate to country-specific developmental problems, needs, and conditions.

Research/Projects:
    Initiating and soliciting support of major projects aimed at the selective transfer of appropriate modern technologies to developing countries.

    Identifying funding sources for the promotion of basic and applied research in areas of scientific interest to developing countries.

    Acting as a liaison between governments and sponsoring organizations in the identification of educational institutions that best satisfy the academic needs of scholars and students from developing countries.

    Developing channels of communication between scholars and scientists in the developing and developed world to facilitate dissemination of information.

    Compiling a comprehensive annotated bibliography of research work dealing with appropriate technology.

    Directing the collection, organization, and dissemination of existing knowledge and information dealing with appropriate technology.

Publications:
> *Approtech*, quarterly journal.
>
> Additional list available upon request.

Montana

Public Interest

National Center for Appropriate Technology
P.O. Box 3838
Butte, Montana 59701
Telephone: (406) 494-4572

Contact:
> Thomas H. Pelletier

Objectives/Goals:
> Nonprofit corporation funded by the U.S. Community Services Administration to promote local programs that develop individual and community self-reliance and to promote the usage of appropriate technologies for specific needs of low-income people.

Research/Projects:
> Provides technical research.
>
> Provides financial assistance to various appropriate technology projects.
>
> Disseminates information and publications related to appropriate technology.

Publications:
> List available upon request.

New Hampshire

Public Interest

Clamshell Alliance
62 Congress Street
Portsmouth, New Hampshire 03801
Telephone: (603) 436-5414

Objectives/Goals:
: A coalition of 30 local organizations opposing construction of a nuclear power plant in Seabrook, New Hampshire.

Research/Projects:
: Information available upon request.

Publications:
: <u>Clamshell Alliance News</u>, bimonthly newsletter.

New Jersey

Academic

Rutgers University
Department of Urban Studies
New Brunswick, New Jersey 09803
Telephone: (201) 932-4006

Contact:
: Michael R. Greenberg, Professor, Environmental Planning

Objectives/Goals:
: To study the impact of man's activities on the environment.

Research/Projects:
: To study the spatial distribution of and temporal changes in cancer mortality.

  To study the correlates of the manufacture and use of toxic substances.

Publications:
: List available upon request.

Health

New Jersey Committee for Occupational Safety and Health (NJCOSH)
80 Central Avenue
Clark, New Jersey 07066
Telephone: (201) 486-4739

Contact:
> Michael McKowne, Chairperson

Objectives/Goals:
> To educate workers and community residents on occupational hazards and the technical, legal, and organizing tactics to improve health and safety conditions.
>
> To provide information about technical services and professional resources for worker health efforts.
>
> To organize union and community support for key health and safety struggles.
>
> To insure that the occupational safety and health administration and other government agencies are fulfilling their responsibilities to New Jersey residents.

Research/Projects:
> Low-level radiation and its health effects.
>
> Incidence of upper respiratory occupational diseases in the United States.
>
> Effects of cutting oils and oil mist on a cohort of 1,400 metal workers.

Publications:
> Monthly newsletter.

New Mexico

Public Interest

Americans for Indian Opportunity
Plaza Del Sol Building
Suite 808
600 Second Avenue, N.W.
Albuquerque, New Mexico 87102
Telephone: (505) 842-0962

Contact:
> La Donna Harris, President; Margaret L. Gover, Project Director

Objectives/Goals:
: To provide technical assistance, expertise, and research to Native American Indian Tribes/individuals in self-government and economic self-sufficiency.

Research/Projects:
: An evaluation of the environmental health impacts of development on Indian communities and the roles of government agencies charged with the responsibilities for various aspects of environmental protection and individual safety.

Publications:
: Indians, Coal and the Big Sky, by Fred and LaDonna Harris, 1974.

    Coal: Black Death for Red Culture, by David Logsdon.

    Real Choices in Indian Resource Development, Reston, Virginia, Conference, 1974.

    Economic Self-Sufficiency and Sovereignty (Pacific Island of Nauru), A Parallel for Indian Nations.

    Indian Resource Projects: A Note on Possible Forms of Taxation, by Stephen Zorn.

    Capital, Technology, and Development, by Harry Magdoff.

    American Indian Economic Development, by A. T. Anderson.

    Indian Irrigation Projects, by Gail Offen.

    Indian Minerals, by Rod Trosper.

    Getting a Fair Deal in Mining Projects, by Stephen Zorn.

    Tourism and Recreation . . . Some Facts, by Tom Loder.

    A Violation of Trust—Federal Mismanagement of Indian Forest Lands, by Rich Nafziger.

    Indian Tribes as Developing Nations, A Question of Power: Indian Control of Indian Resource Development.

130 / SAFETY AND HEALTH CATALOGUE

Hard Choices: Development of Non-Energy Non-Replenishable Resources, AIO Report: Indian Rights and Claims.

Historical Overview of U.S.-Indian Relations and Economic Development, A Chronological Summary.

Alternatives for Development of Agriculture, "Case Studies of Successful Tribal Farms."

What About Irrigation? For Tribal Decision-Makers, AIO Red Paper.

You Don't Have to Be Poor to Be Indian, Maggie Gover.

New York

Academic

Columbia University School of Public Health
East Coast Health Discussion Group
600 W. 168th Street
New York, New York 10032
Telephone: (212) 694-3912

Contact:
    Sally Guttmacher

Objectives/Goals:
    Study and discussion group for critical analysis from a socialist perspective of health-related issues, health care, and the etiology of disease.

Research/Projects:
    Conference on Industrialization and Health, held in Detroit, October 18, 1980.

Publications:
    Health maintenance organization packets of readings: No. 4 Medicine and Ideology; No. 5 Occupational Health; No. 6 Imperialism, Dependency and Health.

Columbia University School of Public Health
Women's Occupational Health Resource Center

60 Haven Avenue, B-1
New York, New York 10032
Telephone: (212) 694-3464

Contact:
    Jeanne M. Stellman, Ph.D., Executive Director

Objectives/Goals:
    A central clearinghouse and communication network on women and occupational health concerns.

Research/Projects:
    Programs designed to develop occupational health awareness in collaboration with unions, management, and other interested organizations.

    Workshops and worker-oriented programs on such topics as personal protective equipment, industrial safety, and the hazards of household work.

    Clearinghouse activities: Collects and disseminates information and publications produced by other organizations concerned with women's occupational health.

    Provides technical assistance in setting up programs and consulting with management.

    Extensive research library and information service.

Publications:
    Newsletter, bimonthly.

    Technical Bulletin, quarterly.

    Facts sheets on specific risks.

    Selected bibliographies and fact packs.

Cornell University
New York State School of Industrial and Labor Relations
(Occupational Safety and Health Programs)
3 East 43rd Street
New York, New York 10017
Telephone: (212) 599-4550

Contact:
    Frank Goldsmith, Director

Objectives/Goals:
    To deliver occupational safety and health educational services to working people in the State of New York. A certificate program in job safety and health is available to trade unionists. Up to 12 Cornell credit hours may be earned in this certificate program. In addition, short-term tailor-made courses for trade unions are available for the training of union personnel, shop stewards, officials, and so on, in job safety and health.

Research/Projects:
    Research has taken place in collective bargaining and OSHA issues. These include arbitration of safety and health issues, questions involving duty of fair representation under the National Labor Relations Act, workers' compensation for occupational diseases, and other contractual issues.

Publications:
    Information available upon request.

Cornell University
Program on Science, Technology and Society
632 Clark Hall
Ithaca, New York 14853
Telephone: (607) 256-3810

Contact:
    Dr. Sheila Jasanoff, Research Associate

Objectives/Goals:
    To bring together scholars from various disciplines to carry out research and teaching on the social and humanistic implications of scientific and technological development.

Research/Projects:
    Cross-national policy making on the environment (with special emphasis on toxic substances control).

    Risk assessment in environmental regulation.

    Compensation for pollution-related injuries.

Publications:
"Compensation for Victims of Toxic Substance Pollution: Allocating Burdens," in <u>Toxic Substances: Decisions and Values III</u>, Report on Conference sponsored by the Technical Information Project, Washington, D.C. (1979).

Hunter College
School of Health Sciences
440 E. 26th Street, Box 611
New York, New York 10010
Telephone: (212) 481-5119 or 481-5120

Contact:
Manuel Gomez or Dr. George J. Kupchik

Objectives/Goals:
To educate and train students at the undergraduate and masters' levels in occupational health and safety.

Research/Projects:
Information available upon request.

Publications:
Information available upon request.

Montefiore Hospital and Medical Center
Department of Social Medicine
111 E. 210th Street
Bronx, New York
Telephone: (212) 920-5508

Contact:
Victor W. Sidel, M.D., Chairperson

Objectives/Goals:
To educate and document the needs of indigent people in the Bronx, specifically, the United States, and the global community.

The occupational and international health divisions are specifically concerned with international environmental and occupational safety and health issues.

134 / SAFETY AND HEALTH CATALOGUE

Research/Projects:
>Information available upon request.

Publications:
>Information available upon request.

Mt. Sinai School of Medicine Environmental Sciences Laboratory
100th Street and 5th Avenue
New York, New York 10029
Telephone: (212) 650-6500

Contact:
>Dr. Irving J. Selikoff

Objectives/Goals:
>Education, training, and research.

Research/Projects:
>Extensive list available upon request.

Publications:
>Extensive list available upon request.

Health

North American Congress on Latin America (NACLA)
151 West 19th Street
9th Floor
New York, New York 10011
Telephone: (212) 989-8890

Contact:
>Helen Shapiro

Objectives/Goals:
>Independent, nonprofit research collective focusing on the political economy of the Americas.

Research/Projects:
>Information available upon request.

Publications:
>Report on the Americas, bimonthly magazine.

>Extensive list available upon request.

NORTH AMERICA / 135

International

United Nations Development Program (UNDP)
One United Nations Plaza
New York, New York 10017
Telephone: (212) 754-1234

Contact:
    Gordon Havord, Director, Division for Program Development, Support, and Evaluation

Objectives/Goals:
    To support projects of technical assistance in developing countries. Each project is designed to form part of a coordinated country or regional program. Drawn up by the governments concerned, these programs are planned for five-year periods and are based on overall national priorities.

Research/Projects:
    Various research projects concerned with occupational safety and health are executed by the International Labor Office, acting as UNDP's executive agent. They include: occupational health; assistance to occupational safety and health; medecine du travail; labor medicine, security, and hygiene; occupational hygiene and safety; training in industrial relations and safety measures; development of central and regional occupational health and industrial hygiene laboratories; Occupational Safety and Health Institute; safety control in steam boiler and pressure vessels equipment; environmental and occupational health training; strengthening of the factory inspectorate in Sind and Punjab; evaluation of toxicity of new technologies of substances used in industry; National Institute of Labor Science; occupational safety and health—factory inspectorate; regional laboratories of industrial hygiene in six regions.

Publications:
    Extensive list available upon request.

Labor

Occupational Safety and Health Training
American Federation of State, County and Municipal Employees

District Council 37
140 Park Place
Room 814
New York, New York 10017
Telephone: (212) 766-1562

Contact:
    Marsha Love

Objectives/Goals:
    To develop research and materials on hazard awareness in high-risk occupations within District Council 37.

Research/Projects:
    Current research areas include: road crews; motor vehicle operators; sewage treatment workers; climbers and pruners (forestry workers/gardeners); laundry workers; stationary firemen.

Publications:
    Fact sheets on heat-stress, asbestos.

    Bibliography sheet.

    <u>On the Road to Safety and Health: A Manual for Road Crews</u>.

Public Interest

American Lung Association/American Thoracic Society
1750 Broadway
New York, New York 10019
Telephone: (212) 245-8000

Contact:
    James A. Swomley, Managing Director

Objectives/Goals:
    The American Lung Association (ALA): A national voluntary health agency conducting educational programs to prevent and control lung disease.

    The American Thoracic Society (ATS): A professional medical and scientific society and the medical section of the American Lung Association.

Research/Projects:
:   ALA: to educate health professionals, patients, special interest groups, and the general community in the following areas: prevention of occupational lung disease; adult lung diseases; pediatric lung diseases; prevention of smoking; improvement of air quality.

    ATS: organized into seven scientific assemblies, including the scientific assembly on environmental and occupational health, concentrating on specialized areas of interest.

Publications:
:   Occupational Lung Diseases: An Introduction, 1979.

    Fact Series on Occupational Lung Disease, 1980.

    ALA Bulletin.

    American Review of Respiratory Disease.

    Basics of RD.

    Clinical Notes on Respiratory Disease.

Asia/North America Communications Center
110 Terrace Blvd. Room 304
New Hyde Park, New York 11040
Telephone: (516) 437-6064

Contact:
:   Thomas P. Fenton

Asia/Pacific
2 Man Wan Road
17-C
Kowloon, Hong Kong
Telephone: 3-035271

Contact:
:   John Sayer/ Christine Vertucci

Objectives/Goals:
:   Hong Kong-based nonprofit Research and Documentation Center. Major objective is to systematically gather,

organize, analyze, and publish information on the U.S. economic, political, and military involvement in Asia.

Research/Projects:
Current Projects: Compilation of descriptive data on all transnational corporations active in the Asia region in collaboration with the United Nations Center on Transnational Corporations. Data include parent and affiliate names; affiliate addresses; line of business; sales; capitalization, assets; historical information; and industry breakdown on all Asian-based affiliates.

Past Projects: Asia in America, prepared for the UNITAR meeting in New Delhi, India, March 11-18, 1980. A statistical summary of U.S. direct investments in Asia. It includes the most current and complete collection of published and unpublished data available from the U.S. Department of Commerce on U.S. direct investments abroad.

Prospective Projects: A major study of health and safety-related issues in Asia. The focus will be on one industry (for example, chemical or asbestos industry) in one particular country as an illustration of conditions in U.S. multinational corporations throughout Asia, to begin Fall 1980.

Publications:
Asia Monitor, quarterly digest of U.S./Asia-related economic news.

America in Asia: Research Guides on United States Economic Activity in Pacific Asia: Vol. I, descriptive listing of sources of information in Asia on U.S. economic involvement in Asia.

America in Asia: A Handbook of Facts and Figures on United States Economic and Military Activity in Pacific Asia: Vol. II, Sourcebook of data on U.S. economic and military power in Asia.

A Survey of Education/Action Resources on Multinational Corporations, an extensive bibliography of resources on multinational corporations.

Center for International Environment Information
300 East 42nd Street
New York, New York 10017
Telephone: (212) 697-3232

Contact:
>Whitman Bassow, Executive Director

Objectives/Goals:
>To increase public understanding in Canada and the United States of international environmental issues.

>To transfer information to policy makers in government and industry in the global community.

Research/Projects:
>International Environment Forum: a membership group meeting four times annually to provide American and Canadian executives and senior foreign environmental officials with information to examine global environmental problems and policies.

>Media Services: the environmental issues guide series provides the news media, as well as concerned professionals, with authoritative information services drawn from government, industry, academic institutions, and environmental organizations.

>World Environment Report, an eight-page newsletter, biweekly, published in North America, focusing exclusively on global environmental issues for policymakers in government, industry, international organizations, and specialists informed about worldwide environmental developments.

Publications:
>List available upon request.

Center for Occupational Hazards
5 Beekman Street
New York, New York 10038
Telephone: (212) 227-6220

Contact:
>Michael McCann

Objectives/Goals:
    Research and information publications on occupational health problems of artists, craftspeople, and related workers, including information on children's art materials.

Research/Projects:
    Art Hazards Information Center, for phone and written inquiries.

    Lectures, workshops.

Publications:
    Art Hazards Newsletter, $10/year.

    <u>Health Hazards Manual for Artists</u>, by M. McCann, $4.00/copy.

    <u>Artist Beware: The Hazards and Precautions in Working with Arts and Crafts Materials</u>, by M. McCann, $16.95.

Coalition of Labor Union Women
15 Union Square
New York, New York 10003
Telephone: (212) 777-5330

Contact:
    Ellen Gurzinski, Executive Director

Objectives/Goals:
    To provide resource and educational materials for working women specifically and working people in general.

    Affirmative action in the workplace.

    Political and legislative action.

    Participation of women in unions.

Research/Projects:
    Maintains reprints and articles of interest and concern to working women, among them, information on occupational safety and health.

Publications:
    <u>Lead: New Perspectives on an Old Problem</u>, booklet.

Council on Economic Priorities
84 Fifth Avenue
New York, New York 10011
Telephone: (212) 691-8550

Objectives/Goals:
    Nonprofit organization to improve the quality of U.S. corporate performance in the critical areas of our social and natural environment.

Research/Projects:
    Occupational safety and health: to examine and evaluate the compliance records of eight major U.S. chemical companies with federal OSHA regulations and federally approved state OSHA programs.

    To assess the OSHA enforcement system relative to the companies examined.

    Jobs and energy: to analyze two contrasting scenarios for energy and economic growth on Long Island, New York.

    Buying power: to analyze the lobbying practices and political influence structures maintained by the major military contracting companies.

    The market and the community: to study public pension fund investment strategies for economic impact and social responsibility based on a case study prepared for Massachusetts.

    Timber management: to examine the scenario for optimal future availability of timber and to discuss the legal and economic constraint on full productivity.

Publications: Recent publications include:
    <u>Conversion Information Center Newsletter</u>, January 1979.

    <u>Mine Control</u>, November 1978.

    <u>Power Politics</u>, October 1978.

    <u>Minding the Corporate Conscience</u>, June 1978.

    <u>The Price of Power/Update</u>, Ronald White, December 1977.

Weapons for the World/Update, Steven Lydenberg, November 1977.

Environmental Steel/Update, Fred Armentrout and James Cannon, September 1977.

Nuclear Plant Performance/Update, Charles Komanoff and Nancy A. Boxer, May 1977.

Shortchanged/Update, Tina Simcich, December 1976.

The Invisible Hand, Gordon Adams, December 1976.

Power Plant Performance, Charles Komanoff, November 1976.

Health Policy Advisory Center (Health PAC)
17 Murray Street
New York, New York

Contact:
    Marilynn Norinsky

Objectives/Goals:
    Independent, nonprofit public interest center, concerned with monitoring and interpreting the health system to change-oriented groups of health workers, consumers, professionals, and students.

Research/Projects:
    Maintains a resource center; offers a network for health activists to discuss, share, and analyze their concerns and strategies with others across the country engaged in similar endeavors.

    Produces and distributes educational materials on social, political, and economic issues in health care.

    Publishes bimonthly *Bulletin*, bringing readers both facts and an understanding of important developments taking place in the health system, including a regular column on occupational health (providing research and analysis on occupational health providers and OSHA policy, and so on).

Publications:
> Bulletin, bimonthly publication; columns on national health policy, women's health, NYC health policy, and occupational and environmental health.
>
> American Health Empire: Power, Profits, and Politics, Vintage Books.
>
> Prognosis Negative: Crisis in the Health Care System, Vintage Books.
>
> Pamphlets, including: The Demise of Public Hospitals, New York City Health Politics, Occupational Health, Health Workers, National Health Insurance, Prepaid Group Practice, Patients' Rights, Mental Health, Women and Health, Primary Care.

Inform, Inc.
25 Broad Street
New York, New York 10004
Telephone: (212) 425-3550

Objectives/Goals:
> Nonprofit research and education organization studying the impact of U.S. corporations on the environment, employees, and consumers.
>
> The occupational safety and health program studies workplace conditions and alternatives for improvement as well as safety and health policy questions.

Research/Projects:
> At Work in Copper: Occupational Health and Safety in Copper Smelting, April 1979; a study of the occupational health and safety conditions in the copper smelting industry.
>
> The effectiveness of governmental regulations in the nine copper smelter states.
>
> Workplace conditions in the lead, zinc, and cadmium smelters.

144 / SAFETY AND HEALTH CATALOGUE

Publications:
    At Work in Copper: Occupational Health and Safety in Copper Smelting, April 1979.

    Questions and Answers: Occupational Health and Safety in Copper Smelting, April 1980.

Institute for Labor Education and Research
853 Broadway
Room 2007
New York, New York 10003
Telephone: (212) 674-3322

Objectives/Goals:
    To teach courses and develop curriculum materials for rank-and-file workers in the United States.

Publications:
    Three slide shows: "Should Energy Cost an Arm and a Leg?"; "Your Job and Your Life"; "Is Affirmative Action Still Alive?"

    What's Wrong with the U.S. Economy? (workers' guide to the economy).

    Anthology on Work for Working People.

Labor Safety and Health Institute
377 Park Avenue
New York, New York 10016
Telephone: (212) 689-8959

Contact:
    Frank Goldsmith, Director

Objectives/Goals:
    To maintain an extensive library on current occupational safety and health information for educational training and research work. Loose-leaf binders contain newspaper clippings from major dailies, research studies, and other related information.

Publications:
: Occupational Safety and Health Handbook.

: Occupational Safety and Health Workbook.

National Audubon Society
950 Third Avenue
New York, New York 10022
Telephone: (212) 832-3200

Contact:
: Bette Tedford

Objectives/Goals:
: To improve environmental quality for all forms of life.

Research/Projects:
: Plans to research areas of hazardous wastes, energy, coastlines and oceans, and water resources.

Publications:
: Available upon request.

Natural Resources Defense Council (NRDC)
122 E. 42nd Street
New York, New York 10017
Telephone: (212) 949-0049

Contact:
: Carol Hine

917 15th Street, N.W.
Washington, D.C. 20005
Telephone: (202) 737-5000

Contact:
: Jacob Scherr, Esq.

Objectives/Goals:
: Nonprofit national legal group that litigates on environmental issues, participates in administrative proceedings, and directs public educational projects.

146 / SAFETY AND HEALTH CATALOGUE

Research/Projects:
> Current projects include: toxic substances, energy, air and water pollution, coastal zone preservation, transportation, international toxics, short wave radiation.

Publications:
> NRDC Newsletter.
>
> AMICUS (monthly publication).

New York Committee for Occupational Safety and Health (NYCOSH)
P.O. Box 3285
Grand Central Station
New York, New York 10017
Telephone: (212) 599-4592

Contact:
> Deborah Ann Nagin, Coordinator

Objectives/Goals:
> To provide technical assistance, worker training, and political action for unions.
>
> To enforce and improve occupational safety and health-related laws in the workplace.

Research/Projects:
> Occupational safety and health training for N.H.S.C. assignees (HEW contract).
>
> Occupational hazards and reproductive rights.
>
> Community perspectives on job health.

Publications:
> Bimonthly newsletter.
>
> Fact sheets.
>
> Reproductive hazards course materials.

Scientists' Institute for Public Information (SIPI)
355 Lexington Avenue

New York, New York 10017
Telephone: (212) 661-9110

Contact:
    Alan McGowan, President

Objectives/Goals:
    Nonprofit organization communicating scientific information to nonspecialists.

Research/Projects:
    Maintains data bank with over 2,000 names of scientists filed and cross-referenced by disciplines.

    Arranges interviews with specialists for newspapers, magazines, and journals through the Scientist-Reporter Liaison Program and the Media Outreach Program.

    Current research includes: evaluation of computer usage in law enforcement; the use of methane as a fuel source; evaluation of the effects of government regulation on the oil industry; public information project with the Tennessee Valley Authority to organize and conduct citizen participation programs on nuclear waste storage.

Publications:
    SIPIscope Newsletter; includes information on SIPI activities and serves as a discussion forum.

    Environment Magazine, published in cooperation with the Helen Dwight Reid Education Foundation, covering current ecological issues.

    List available upon request.

Shop Talk Productions
155 West 72nd Street, Room 402
New York, New York 10023
Telephone: (212) 580-1881

Contact:
    Nick Egleson or Bonnie Bellow

Objectives/Goals:
: To produce audiovisual materials that educate workers on occupational safety and health hazards.

Research/Projects:
: Presentations on: asbestos, benzene, acrylonitrile, arsenic, textile industry hazards, and worker's rights through their unions and OSHA to work for better conditions on the job.

Sierra Club
International Office
800 Second Avenue
New York, New York 10017
Telephone: (212) 867-0080

U.S. Office
530 Bush Street
San Francisco, California 94108
Telephone: (415) 981-8634

Contact:
: (in New York) Program Associate; (in San Francisco) Carl Pope, Labor Liaison, Conservation Department

Objectives/Goals:
: To explore, enjoy, and protect the world's natural heritage.

Research/Projects:
: Industrial safety standards relating to issues of environmental health.

  Public education and lobbying activities.

  Symposia in cooperation with European organizations on the harmonization of National Toxics Regulations.

  Nuclear waste management task force.

Publications:
: Fact sheets available on various aspects of toxics regulation.

  List available upon request.

NORTH AMERICA / 149

Religious

National Council of the Churches of Christ: USA, Fifth Commission
475 Riverside Drive, No. 866
New York, New York 10027
Telephone: (212) 870-2915

Contact:
    Jovelino Ramos

Objectives/Goals:
    To monitor the National Council of Churches to make it attentive and responsive to justice and liberation causes and Third World concerns.

Research/Projects:
    Consultation development (1974).

    Ongoing policy statements on immigration.

Publications:
    Fifth Commission Newsletter.

North Carolina

Public Interest

Institute for Southern Studies
P.O. Box 531
Durham, North Carolina 27702
Telephone: (919) 688-8167

Objectives/Goals:
    Nonprofit educational organization providing strategic information to various social-change organizations, labor unions, and health and safety groups.

Research/Projects:
    Information available upon request.

Publications:
    Southern Exposure ($10 per year); quarterly journal featuring topically organized issues including resources on occupational health and labor organizing in the South.

Sick for Justice and Here Comes a Wind ($4.50); other issues on the nuclear power industry, women in the South, and labor history during the 1930s and 1940s.

North Carolina Occupational Safety and Health Project (NCOSH)
P.O. Box 2514
Durham, North Carolina 27705
Telephone: (919) 286-2276

Contact:
Alan Weiner, Lee Guion, David Austin

Objectives/Goals:
Independent nonprofit group of industrial hygienists, nurses, lawyers, public health workers, and students working to provide the following services: courses and workshops for unions and workers on occupational safety and health issues; research and analysis on specific health hazards; medical and legal referrals; health screening clinics for occupational disease.

Research/Projects:
Chemical analysis of toxic substances.

Screening clinics for rubber and textile workers.

Publications:
Slide show, "We Pay with Our Lives" (profiles of occupational safety and health problems in the workplace).

NCOSH Reports, newsletter.

Cancer and White Lung (a report on the plight of North Carolina's asbestos workers).

Ohio

Academic

The Ohio State University
Labor Education and Research Service
1810 College Road
Columbus, Ohio 43210
Telephone: (614) 422-8157

Contact:
: C. J. Slanicka, Director

Objectives/Goals:
: A state-supported continuing education program devoted to: developing teaching materials for occupational safety and health training; conducting training in occupational safety and health for employees and union representatives; providing technical assistance to workers and union representatives.

Research/Projects:
: Hazard Recognition Training Program: Health and safety courses designed to provide workers with the knowledge and skills needed to recognize and evaluate workplace hazards and, then, to provide workers with methods for acting upon their newly gained knowledge and skills in order to correct workplace conditions.

    Conducts conferences, seminars, and short courses.

    Plans, designs, and conducts research activities for labor organizations throughout the state.

Publications:
: Information available upon request.

Public Interest

N.E. Ohio Committee for Occupational Safety and Health
1793 Wilton Road
Cleveland Heights, Ohio 44118
Telephone: (216) 932-9344

Contact:
: Sherry Baron, Joe Castorina, Pat Melaine

Objectives/Goals:
: Organization of labor unions in N.E. Ohio area.

    To share information and resources concerning occupational health and safety issues.

Research/Projects:
: Information available upon request.

Publications:
: Information available upon request.

Oregon

Labor/Academic

Pacific Northwest Labor College
Box 25
Marylhurst, Oregon 97036
Telephone: (503) 245-1315

Contact:
: John Lund, Director, Department of Safety and Health

Objectives/Goals:
: Labor-owned and -operated worker education and labor research institute.

  Instruction and training in theory, arts, and practicum, including those of citizenship, pertinent to the administration, programs, and functions of unions and employee associations.

  Research into questions, problems, subjects, and issues of concern and interest to workers and their organizations.

  Seminars, conferences, publications, and other public services and information dissemination activities of importance to workers and the general public.

  Administration of funds for workers' educational programs, including programs conducted jointly with other institutions and organizations, grants for labor research awarded to the Pacific Northwest Labor College or by the college to other institutions and individuals, and scholarships for students of Pacific Northwest Labor College.

  Lifelong learning programs for officials and members of organized labor, retired persons, and interested youth.

Research/Projects:
: Worker education in safety and health.

Resource library on environmental and occupational health issues.

Research and technical assistance for unions and workers on health and safety issues, hazard recognition, and the use of occupational health standards.

Publications:
Research and Education, monthly newsletter.

Shop Stewards Safety and Health Manual.

Academic course materials.

Pennsylvania

Academic

Pennsylvania State University College of Human Development
University Park, Pennsylvania 16802
Telephone: (814) 865-3447

Contact:
Vilma R. Hunt, Associate Professor of Environmental Health

Objectives/Goals:
Teaching occupational and environmental health to undergraduate and graduate students.

Research/Projects:
Occupational hazards.

Women and work hazards.

Publications:
Information available upon request.

Pennsylvania State University
Materials Research Laboratory
University Park, Pennsylvania 16802
Telephone: (814) 865-3422

Contact:
Professor Rustum Roy

Objectives/Goals:
Education and research.

Research/Projects:
Radioactive waste management.

Publications:
Information available upon request.

University of Pittsburgh Graduate School of Public Health
Department of Industrial and Environmental Health
130 DeSoto Street
Pittsburgh, Pennsylvania 15261
Telephone: (412) 624-3045

Contact:
Thomas F. Mancuso, M.D., Research Professor

Objectives/Goals:
Determination of long-term delayed effects of the microchemical environment and of low-level ionizing radiation.

Research/Projects:
Cancer studies on migrant populations.

Epidemiological prospective worker studies relative to asbestos, beryllium, chromate, betanaphthylamine, and benzidine, rubber, and rayon industries and radiation in atomic energy industries. Emphasis of all the studies has been on environmental cancer.

Publications:
Reprints available upon request.

Labor

United Steelworkers of America
Safety and Health Department
Five Gateway Center
Pittsburgh, Pennsylvania 15222
Telephone: (412) 562-2581

Contact:
Adolph E. Schwartz, Director, Safety and Health Department

Objectives/Goals:
: To reduce and eliminate work hazards through a series of worker education programs in hazard recognition.

Research/Projects:
: Worker education programs in both general safety and health principles and specific high-hazard industries.

    Preparation of worker education materials in occupational safety and health.

Publications:
: <u>Occupational Safety and Health Manual</u>.

    <u>Proposed Safety and Health Contract Language</u>.

    <u>Health Hazards in the Aluminum Industry</u>.

Public Interest

American Friends Service Committee
1501 Cherry Street
Philadelphia, Pennsylvania 19102
Telephone: (215) 241-7000

Objectives/Goals:
: A Quaker organization engaged in programs of service, education, and social change to promote global economic, political, civil, social, and human rights.

Research/Projects:
: Inequalities in health care and other basic services resulting from poverty, sex, and race.

    The status of native Americans.

    Discrimination against blacks, Hispanics, other minorities, and women in critical areas such as housing, education, and employment.

    Additional information available upon request.

Publications:
: List available upon request.

156 / SAFETY AND HEALTH CATALOGUE

Miners Clinics, Inc.
1260 Martin Avenue
New Kensington, Pennsylvania 15068
Telephone: (412) 339-6641

Contact:
    Donald J. Conwell, Executive Director

Objectives/Goals:
    To operate and maintain several comprehensive out-patient clinics for miners.

Research/Projects:
    Current medical services include: primary and specialty medical care; ancillary services: laboratory, x-ray, podiatry, optometry, and physical therapy; special programs include: family planning, home health, and a coal workers respiratory disease program.

Publications:
    Brochures available relating to specific programs and overall clinic operations.

Philadelphia Area Project on Occupational Safety and Health (PhilaPOSH)
1321 Arch Street
Room 201
Philadelphia, Pennsylvania 19107
Telephone: (215) 568-5108

Contact:
    Rick Engler

Objectives/Goals:
    Nonprofit coalition of local unions, workers, and health and legal professionals providing training on occupational health issues to Delaware Valley local unions and labor groups. Education sessions teach the skills and information needed to build and maintain an effective health and safety effort.

Research/Projects:
    Technical support: research on hazards, referrals for medical and legal services, assistance with writing contract language and filing OSHA complaints.

Resource Center and library for workers with various publications available (write for publication list).

Publications:
Safer Times, Delaware Valley Health and Safety Newsletter.

Oil Refinery Health and Safety Hazards (guide to refinery hazards).

Health Technical Committee Handbook.

Fact Sheets on worker's compensation: ventilation, noise, radiation, reproductive hazards, asbestos, trichloroethylene, heat, and others.

Puerto Rico

Public Interest

Legal Services of Puerto Rico, Inc.
P.O. Box 987
Aguadilla, Puerto Rico 00603
Telephone: (809) 891-5330

Contact:
Manuel J. Vera Vera, Staff Attorney

Objectives/Goals:
To provide legal services to the indigenous population.

Research/Projects:
Class-action litigation resulting from an asbestos cement housing project. Suit demands the adoption and application of the local Occupational Safety and Health Act.

Working for the application of Environmental Protection Agency standards for occupational and environmental exposure to asbestos in relation to asbestos cement materials.

South Carolina

Public Interest

Southerners for Economic Justice
P.O. Box 3084
Greenville, South Carolina 29602

1931 Laurel Avenue
Knoxville, Tennessee 37916
Telephone: (615) 637-8019

Contact:
Jim Sessions, Mike Russel, Syl Sampson

Objectives/Goals:
To promote economic justice concerns in the South.

To advocate and monitor workers' rights in the Southern workplace.

To support workers' rights in collective bargaining.

Research/Projects:
Research and monitor reports on: workers' compensation, antilabor forces in the South, and affirmative action.

Work with Southern institutions (civil rights organizations, clergy, and so on) to help neutralize antiunion sentiment in the South and to promote basic civil rights in the workplace.

Publications:
Greenville Report.

Workers' Compensation: Who Really Gets Compensated?

Fair Measure, newsletter, reporting Southern labor developments and other social justice concerns.

Various articles on religion and labor in the South.

Tennessee

Public Interest

Southerners for Economic Justice
1931 Laurel Avenue
Knoxville, Tennessee 37916
Telephone: (615) 637-8019

P.O. Box 3084
Greenville, South Carolina 29602

Contact:
    Jim Sessions, Mike Russel, Syl Sampson

Objectives/Goals
    To promote economic justice concerns in the South.

    To advocate and monitor workers' rights in the Southern workplace.

    To support workers' rights in collective bargaining.

Research/Projects:
    Research and monitor reports on: workers' compensation, antilabor forces in the South, and affirmative action.

    Work with Southern institutions (civil rights organizations, clergy, and so on) to help neutralize antiunion sentiment in the South and to promote basic civil rights in the workplace.

Publications:
    Greenville Report.

    Workers' Compensation: Who Really Gets Compensated?

    Fair Measure, newsletter, reporting Southern labor developments and other social justice concerns.

    Various articles on religion and labor in the South.

Tennessee Committee on Occupational Safety and Health (TNCOSH)
Center for Health Services
Station 17

160 / SAFETY AND HEALTH CATALOGUE

Vanderbilt Medical School
Nashville, Tennessee 37232
Telephone: (615) 322-4773

Contact:
    Jamie Cohen

Objectives/Goals:
    To provide educational programs, technical assistance, and direct support for workers who face unsafe conditions in their workplace.

    To hold the Tennessee Occupational Safety and Health Administration (TOSHA) accountable for the enforcement of health and safety laws.

Research/Projects:
    Summer internship programs for students in occupational safety and health. Students work for a ten-week period with workers and workers' representatives, helping them identify and remedy hazards in the workplace. The purpose is to sensitize students to workers' concerns and to train them to work with health and safety problems.

    Workshop for workers on teaching health and safety information to their locals and other workers.

Publications:
    List available upon request.

Texas

Academic

Texas A&M University
Systems Building, Room 312
College Station, Texas 77843
Telephone: (713) 845-4417

Contact:
    Ray Frisbie, Integrated Pest Management Coordinator

Objectives/Goals:
    Teaching, research, and extension education programs.

Research studies to minimize environmental pollution by pesticides while maintaining suitable farming profits and human health.

Research/Projects:
Specialization of integrated pest management (PRM), a multi-disciplinary approach to reducing pest (insects, weeds, diseases, and so on) damage to crops, man, and animals.

Publications:
List available upon request.

University of Texas School of Public Health
Texas Occupational Safety and Health Educational Resource Center
P.O. Box 21186
Houston, Texas 77025
Telephone: (713) 792-7450

Contact:
Dr. M. M. Key, Director

Objectives/Goals:
Postgraduate and continuing education in occupational safety and health.

Research/Projects:
Cytogenetic monitoring.

Indoor air pollutants.

Occupational epidemiology.

Publications:
None.

Utah

Academic

University of Utah
Building 112
Salt Lake City, Utah 84112
Telephone: (801) 581-7107

Contact:
> Jeffrey S. Lee, Ph.D., C.I.H.

Objectives/Goals:
> Graduate education, research, and services in occupational health.

Research/Projects:
> Health effects of oil shale and other energy industries.
>
> Health effects of heavy metals including chromium and lead.
>
> Studies of asbestos and fissous zeolytes.
>
> Studies of mining operations.

Publications:
> Information available upon request.

Washington

Public Interest

SER/Jobs for Progress
9826 14th Avenue, S.W.
Seattle, Washington 98106
Telephone: (206) 764-4350

Contact:
> Adrian Moroles, OSHA Program Director

Objectives/Goals:
> To develop bilingual (English/Spanish) occupational safety and health materials and programs in high-risk industries of Region X (Alaska, Idaho, Oregon, Washington) where there are large numbers of Hispanic workers.

Research/Projects:
> To develop media programs to inform farmworkers of the hazards of pesticides, worker rights and responsibilities, and health hazard recognition.
>
> To develop OSHA program for workers in the food processing industry to inform workers of OSHA standards; hazard recognition; and hazard-chemical and noise information.

Publications:
    Orchard Ladder Safety Program (bilingual, English/Spanish; audiovisual slides and cassette).

    Como Cuidar de su Espalda (back care booklet in Spanish).

Washington Occupational Health Resource Center
Box 18371
Seattle, Washington 98118
Telephone: (206) 762-7288

Contact:
    Vicki Laden

Objectives/Goals:
    Presently forming and defining its objectives.

Research/Projects:
    Studying health hazards of radiation.

    Researching pesticides.

    Working with local unions and shipyard workers.

Wisconsin

Academic

University of Wisconsin (Extension)
School for Workers
701 Park-Regent Medical Building
One South Park Street
Madison, Wisconsin 53706
Telephone: (608) 262-2111

Contact:
    George Hagglund, Professor, Labor Education and Supervisor, Occupational Safety and Health Training Project

Objectives/Goals:
    To provide training programs and technical assistance and to produce educational materials for local union officers, full-time staff, and workers.

164 / SAFETY AND HEALTH CATALOGUE

Research/Projects:
> Programs concentrating on union systems for dealing with workplace problems, ergonomics, and workplace conditions resulting in job stress.

Publications:
> Training films for union officers in the OSHAct, the industrial hygiene inspection procedure, and the responsibilities of a union safety and health committee.
>
> Slide-tape units covering topics such as ergonomics, noise hazards, dermatitis, planning for a better work environment, and chemical hazards.
>
> Printed brochures aimed at educating workers and trade unionists on health and safety issues.

SOUTH AMERICA

Brazil

Academic

University of São Paulo
School of Public Health
Department of Environmental Health
Occupational Health Branch
Avenida Dr. Arnaldo, 715
São Paulo, Brazil
Telephone: (011) 280-3233, branch 35

Contact:
> Professor Diego Pupo-Nogueira, M.D.

Objectives/Goals:
> Postgraduate teaching in occupational health.
>
> Research on occupational health.
>
> Community advisory service on occupational health care.

Research/Projects:
> Organization of medical services in the workplace.

Epidemiological study of pneumoconiosis.

Epidemiology of sick-absenteeism.

Levels of lead, chlorinated pesticides, and PCBs in the "normal" population of São Paulo.

New and simple methods of analysis for cadmium and manganese (under the auspices of the World Health Organization).

Energy expenditure of automotive workers.

Occupational health and small industries.

Study of the small industries in the city of Americana.

Effects of noise on metallurgical workers.

Exposure to benzene in the shoe industry.

Chrome lesions in galvanoplasty workers.

Publications:
   Revista de Saude Publica (quarterly).

   Didactic collection of slides and transparencies for overhead projector.

Government

Fundacentro, Fundacão Jorge Duprat Figueiredo de Seguranca e Medicina do Trabalho
Al. Barao de Limeira
539 01202 São Paulo, Brazil

Objectives/Goals:
   Created by Brazilian Federal Law, this foundation is supervised by the ministry of labor to provide the following occupational safety and health services: improved workplace conditions; work accident prevention; research and training courses for workers on industrial hygiene and occupational health and safety.

Research/Projects:
> Educational Branch: to coordinate and realize educational activities in the occupational safety and health fields; to train instructors to administer courses to workers, trade union leaders, and masters agents; to promote regular specialized courses on: occupational safety engineering, occupational medicine, occupational nursing, occupational safety supervision.
>
> Library and information service branch: Information library department open to the public to disseminate information on occupational hygiene, safety, and health. National Center of CIS-OIT: to collect, produce, and distribute documentation concerning occupational safety.
>
> Advisory Planning Branch: to develop research and projects related to occupational injuries. To plan nationwide activities on accident prevention and occupational diseases.
>
> Annual Event: Congresso Nacional de Prevenção de Acidentes: Conpat.

Publications:
> List available upon request.

Chile

Government

Instituto Nacional de Salud Publica
Ministerio de Salud, Chile
Departmento de Salud Ocupacional y Contaminación Atmosferica
Santo Domingo 2398
Santiago, Chile
Telephone: 83-84

Contact:
> Dr. Hernan Oyanguren, M.D., Head, Subdepartment of Occupational Medicine

Objectives/Goals:
> To provide teaching, training, research, and consultation on occupational health care issues.

Research/Projects:
> Current research topics: physical and heat load; pneumoconiosis (silicosis, asbestosis, coal mines, pneumoconiosis, siderosis); toxicology (Pb, Hg, As, solvents); noise; vibration; chronic bronchitis; ergonomy.

Publications:
> Reprints of publications in Revista Medica de Chile, and others.

Colombia

Academic

University of Antioduis
School of Public Health
Apartado Aereo 51922
Medellin, Colombia

Contact:
> Dr. Samuel Henao

Objectives/Goals:
> Teaching, research, and advising in occupational health field.

Research/Projects:
> Investigation on lead contamination in a battery factory.

Publications:
> None at present.

Government

Servicio Seccional Salud
Apartado Aereo 55356
Medellin, Colombia

Contact:
> Gustavo Molina, Chief, Occupational Health Section

Objectives/Goals:
> Prevention of work injuries and occupational illnesses among workers in the Medellin region.

168 / SAFETY AND HEALTH CATALOGUE

Education and training of health personnel in occupational health.

Research/Projects:
Occupational lung diseases among coal miners.

Regional program to curtail pesticide usage in Medellin.

Publications:
Information available upon request.

Peru

International

International Labor Office (ILO)
Regional Department of Americas and Caribbean
Apartado Postal 3638
Lima 100, Peru
Telephone: 40-4850 Lima

Contact:
Bernardo Bedrikow, M.D., M.P.H. Regional Advisor, Occupational Safety and Health

Objectives/Goals:
Health and safety work pertaining to the aims of the ILO. (See Appendix J for more detail on the ILO.)

Research/Projects:
Pertaining to the aims of the ILO.

Publications:
List available upon request.

Pan American Health Organization (PAHO)
Pan American Center for Sanitary Engineering and Environmental Sciences
Casilla 4337
Lima, Peru
Telephone: 35-4135; Telex: 21052

Contact:
Ricardo Haddad

Objectives/Goals:
A specialized PAHO Center dealing with environmental problems, including those related to the work environment.

Assistance to countries in the Americas for the development of public health programs.

Research/Projects:
Special research projects conducted in Chile include: survey of conditions in the Chilean industry; effects of manganese; damage due to asbestos; lead exposure in printing shops; solvent exposure; mercury exposure; carbon monoxide in tunnels excavation.

Publications:
<u>Introduccion a la Higiene Industrial</u>, J. J. Bloomfield and Ricardo Haddad.

Over 100 publications on subjects related to health and safety; list available upon request.

Venezuela

Academic

University of Venezuela
Res. "Carenero" No. 3A
AVDA "La Colina" No. 3A
Urb. Terrazas de Sta. Inez
Caracas, 1080 Venezuela
Telephone: (02) 922170

Contact:
Nancy E. Cezar, M.D.

Objectives/Goals:
Physician in shipyards and construction industries.

Teaching at the University of Venezuela, School of Medicine.

Research/Projects:
Information available upon request.

Publications:
Information available upon request.

SOUTH PACIFIC

Indonesia

Academic

Padjadjaram University
Institute of Ecology
1n. Banda 40
Bamdung, Indonesia
Telephone: 022-50901

Contact:
    Professor Dr. Ir. Otto Soemarwoto, Director

Objectives/Goals:
    To advance the science of the ecology of development.

    To educate and train scientists and specialists in the field of the ecology of development.

    To disseminate the knowledge and skills to decision makers, planners, managers, and general public.

Research/Projects:
    Processing and evaluation of data on microcystis bloom.

    Ecology of home garden, village, and rural ecosystem: physical and biological aspects: structure, functioning—biological cycle, productivity and energetics, economic and social aspects.

    Impact of Development: physical and biological aspects, economic and social aspects, health aspects, rural-urban relationships.

    Nature Conservation: conceptual framework, criteria and listing of ecosystems and other features to be preserved, specific studies.

    Environmental impact assessment.

    Statistical and experimental design.

    Stimulation and modeling.

Publications:
>"Management of the Human Environment and National Development," Ecology and Development, 1973.

>"Metals and Chlorinated Pesticides in Samples of Fish, Sawah-duck eggs, Crustaceans and Molluscs Collected in West and Central Java, Indonesia," Ecology and Development, October 1974.

>"Seminar on the Management of Water Resources" (Summary Report), Ecology and Development, January 1975.

>"Dense Forest Areas in the Citarum River Basin" (Summary), Ecology and Development, May 1976.

>"Seminar on the Management of Water Resources" (Abstract), Ecology and Development, July 1977.

Government

National Centre for Industrial Hygiene, Occupational Health, and Safety
JLN ACHMAD YAR1 6g-71
Jakarta, Indonesia
Telephone: 412114

Contact:
>Dr. Suma Mur P.K., Chief of the National Centre

Objectives/Goals:
>To guide the development of industrial hygiene and occupational health and safety in Indonesia by training and education programs, informational activities, research work, standards development, services.

Research/Projects:
>Laboratory construction in 11 provinces.

>Standards development on physical as well as chemical factors.

>Training and certification of physicians engaged in occupational health programs.

Occupational nutrition projects.

Productivity improvement in women workers.

Publications:
Indonesian Journal of Industrial Hygiene, occupational Health and Safety (with some topics in English).

Audiovisuals in Indonesian.

New Zealand

Academic

*University of Auckland
Department of Community Health*
Division of Environmental Health and Safety
Auckland, New Zealand
Telephone: 795780; Telex: N.Z.

Contact:
Tord Kjellstrom

Objectives/Goals:
Research on health effects of the environment, particularly epidemiological research.

Occupational safety and health: main emphasis on metal toxicity.

Assistance to New Zealand unions with information on industrial hazards.

Research/Projects:
Cadmium toxicity in factory workers.

Lead toxicity in factory workers.

Epidemiology of occupational cancer.

Exposure register of asbestos-exposed workers.

Publications:
Research reports and bibliography on request.

Union code of practice for asbestos work.

Government

Occupational Health and W.H.O. Seamen's Centre
CRNR. Quay St. and French St.
Auckland 1, New Zealand
Telephone: Auckland 775620

Contact:
    Dr. James Fren, Port Health Officer, Auckland, and Director, W.H.O. Seamen's Centre

Objectives/Goals:
    Provision of occupational health services to waterside workers.

    Provision of medical services to seamen: base for port health officers, shipping inspectors (of health), St. John's Ambulance Brigade (Waterfront Division of Auckland Brigade), and St. John's Ambulance Association Officers; and base for public health nurses visiting factories in Auckland Waterfront area.

Research/Projects:
    Prospective review of diseases of seamen.

Publications:
    None.

Private/Industrial

New Zealand Health Nurses Association Incorporated
James Hardie & Co., Ltd.
P.O. Box 12070
Penrose
Auckland, New Zealand
Telephone: 599-919 Auckland; Telex: 2712

Contact:
    Susan Greenstreet, National Secretary/Treasurer

Objectives/Goals:
    To promote health and safety in industry.

    To promote education for occupational health nurses.

    To promote community health programs.

Research/Projects:
> Diploma and certification courses for occupational health nurses.
>
> Establish scholarship fund.
>
> Organize annual national conferences.
>
> Maintain reference library.

Publications:
> Currently compiling book of guidelines for nurses and management.

Public Interest

The Australian and New Zealand Society of Occupational Medicine: New Zealand Branch
Medical Directorate
Defense Headquarters
Private BAC
Wellington, New Zealand
Telephone: WN 726-499, x 670

Contact:
> L. J. Thompson, Secretary

Objectives/Goals:
> To promote occupational health and to advance the knowledge, practice, and standing of occupational medicine.

Research/Projects:
> Information available upon request.

Publications:
> List available upon request.

Philippines

Government

Fertilizer and Pesticide Authority (FPA)
Raha Sulayman Building, 6th Floor

Benavidez Street
Makati
Metro Manila, Philippines
Telephone: 85-50-01 to 03

Contact:
  Ricardo T. Deang, Chief, Pesticide Technical Services

Objectives/Goals:
  Common to Fertilizers, Pesticides, and Other Agricultural Chemicals: to conduct an information campaign regarding the sale and effective use of these products; to promote and coordinate all fertilizer and pesticides research in cooperation with the Philippine Council for Agriculture and Resources Research and other appropriate agencies to ensure scientific pest control in the public interest, safety in the use and handling of pesticides, higher standards and quality of products and better application methods; to call upon any department, bureau, office, agency, instrumentality of the government, including government-owned or -controlled corporations, or any officer or employee thereof and on the private sector, for such information or assistance as it may need in the exercise of its powers and in the performance of its functions and duties; to promulgate rules and regulations for the registration and licensing of handlers of these products, collect fees pertaining thereto, as well as the renewal, suspension, revocations or cancellation of such registration or licenses and such other rules and regulations as may be necessary to implement this Decree; to establish and impose appropriate penalties on handlers of these products for violations of any rules and regulations established by the FPA; to institute proceedings against any person violating any provisions of this Decree and/or such rules and regulations as may be promulgated to implement the provisions of this Decree after due notice and hearing; to delegate such selected privileges, powers, or authority as may be allowed by law to corporations, cooperatives, associations, or individuals as may presently exist or be organized to assist the FPA in carrying out its functions; and to do any and all acts not contrary to law or existing decrees and regulations as may be necessary to carry out the functions of the FPA.

  Specifics on Pesticides and Other Agricultural Chemicals: to determine specific use or manners of use for each pesticide or pesticide formulation; to establish and enforce tolerance levels and good agricultural practice for use of pesticides in raw agricultural commodities; to restrict or ban the use of any pesticide or the formulation of certain pesticides in specific areas or during certain periods upon evidence that the pesticide is an imminent hazard, has caused, or is causing widespread serious damage to crops, fish or livestock, or to public health and the environment; to prevent the importation of agricultural commodities containing pesticide residues above the accepted tolerance levels and to regulate the exportation of agricultural products containing pesticide residue above accepted tolerance levels; to inspect the establishment and premises of pesticide handlers to ensure that industrial health and safety rules and antipollution regulations are followed; to enter and inspect farmers' fields to ensure that only the recommended pesticides are used in specific crops in accordance with good agricultural practice; to require if necessary of every handler of these products, the submission to the FPA of a report stating the quantity, value of each kind of product exported, imported, manufactured, produced, formulated, repacked, stored, delivered, distributed, or sold; should there be any extraordinary and unreasonable increase in prices, or a severe shortage in supply of pesticides, or imminent dangers or either occurrence, the FPA is empowered to impose such control as may be necessary in the public interest including but not limited to such restrictions and controls as the imposition of price ceilings, controls on inventories, distribution, and transport and tax-free importations of such pesticides or raw materials thereof as may be in short supply.

Research/Projects:

National Pesticide Safety Program: An agromedical safety program started September 1978 geared toward training on safe handling and use of pesticides and management of pesticide poisoning cases. To date, March 1980, 20 sessions have been conducted addressed to medical doctors, nurses, agricultural technicians, agropesticide dealers, students, plantation/institutional workers and representatives of former organizations, among others. Cooperating agencies include Ministries of Health, Human Settlement, and Agriculture and the National Food and Agriculture Council.

Residue Monitoring: Market-basket sampling of agricultural products for residue analysis, residue monitoring of lakes and water bodies/sources. The project is closely linked with the Pesticide Analytical Laboratories of the Bureau of Plant Industry and the National Crop Protection Center.

Occupational Health and Safety on Manufacturing and Formulation Plants: Improvement/implementation of the standards for occupational health and safety; closely linked with the Ministry of Labor and the chemical companies addressed to the workers and the management. Implemented through dialogues, training sessions, and inspection.

ARSAP Agropesticide Program: A project of the UN Economic and Social Commission for Asia and the Pacific with FPA as the National Coordinating Agency for the Philippines. The program is geared toward upgrading the knowhow of agropesticide dealers on pesticide safety.

Prospective Projects: Development of indigenous source of pesticides that are not of petrochemical origin.

Publications:
Restricted Pesticides in the Philippines.

Safety Tips on Pesticide Handling.

FPA News Bulletin.

"FPA Story" (audiovisual).

Pesticide Poisoning Recognition, Epidemiology and Management (audiovisual series made by the University of Miami School of Medicine).

National Environmental Protection Council (NEPC)
PHCA Building
East Avenue
Quezon City, Philippines
Telephone: 980421, ext. 2634

Contact:
Veronica Villavicencio, Deputy Executive Director

Objectives/Goals:
> To rationalize the functions of government agencies charged with environmental protection.
>
> To formulate policies and issue guidelines for the establishment of environmental quality standards and environmental impact assessments.
>
> To undertake a comprehensive and continuing research program for environmental protection.
>
> To recommend new environmental legislation or amendments to existing law.

Research/Projects:
> Current research includes: coastal zone management; uptake of air pollutants by Philippine plants; mine tailing pollution control management; soil erosion control management; distribution patterns of chemical contaminants and potentially toxic substances; situation profiles for Philippine environmental quality report.

Publications: Current publications including audiovisuals and periodicals:

> Environmental Facts and Features Series 1, 2, 3, 4.
>
> Primer on Environmental Protection.
>
> Primer on Philippine Environmental Decrees.
>
> Reflections: the Philippine Environment into the 21st Century.
>
> Primer on Tax Incentives.
>
> Presidential Decrees Nos. 1151, 1152, 1121.
>
> Primer on National Environmental Protection Council.
>
> Environmental Law: Volumes I and II.
>
> State of the Philippine Environment 1977.
>
> Philippine Environment: Concerns and Issues Digest, 1978.

Public Interest

Asian Science Writers Association
P.O. Box 1843
Manila, Philippines
Telephone: 505026/591478; Telex: PRESSASIA Manila

Contact:
 Adlai J. Amor, Founding Chairman

Objectives/Goals:
 To pool, exchange, and further scientific knowledge among Asian journalists.

 To promote standards and excellence in scientific writing.

 To foster the dissemination of accurate information regarding science through all media normally devoted to the public in Asia.

Research/Projects:
 Annual awards for excellence in reporting about pediatrics.

Publications:
 Newsletter, "The Yenri Principle."

USSR

Government

Institute of Industrial Hygiene and Occupational Diseases
105275 Moscow
Pr. Budennogo 31
USSR
Telephone: 365-02-09; Telex: MOSDVA, "Profgigiena"

Contact:
 Dr. N. F. Izmerov, Director of Institute

Objectives/Goals:
 Principal function is the comprehensive study of all factors of industrial development and the search for methods and measures of preventing their harmful effects on health.

Research/Projects:
   Current research includes: Etiology, pathenogenesis, diagnosis, prevention, and treatment of dust diseases of the respiratory organs; vibration sickness; diseases from noise exposure; diseases of toxic-chemical etiology.

Publications:
   "Work and Health in the Developed Socialist Community," edited by N. F. Izmerov, Medicine (Moscow, 1979).

   "Toxicology of the New Chemical Industrial Substances," edited by N. F. Izmerov, Medicine (Moscow).

   Actual Problems of the Industrial Microclimate, edited by N. F. Izmerov (Moscow, 1978).

   Control of Air Pollution in the USSR, N. F. Izmerov (Geneva, 1973).

   The Journal: Industrial Hygiene and Occupational Medicine.

BIBLIOGRAPHY

Adler, John H. Capital Movements and Economic Development. London/New York: Macmillan/St. Martin Press, 1967.

Alston, Philip. "International Regulation of Toxic Chemicals." Ecology Law Quarterly 7, no. 2 (1975):397-456.

American Center for the Quality of Worklife. Industrial Democracy in Europe. Washington, D.C., 1978.

Anderson, Charles W. "The Changing International Environment of Development in Latin America in the 1970's." Inter-American Economic Affairs 24, no. 2 (Autumn 1970):65-70.

Ashford, Nicholas. Crisis in the Workplace: Occupational Disease and Injuries. Cambridge, Mass.: MIT Press, 1976.

Baker, J. C., and J. K. Ryan, Jr. "Multinational Corporation Investment in Less Developed Countries: Reducing Risk." Nebraska Journal of Economic Business 18, no. 1 (Winter 1979):61-69.

Barkin, Solomon et al., eds. International Labor. New York: Harper & Row, 1967.

The bibliography is a list of additional references on international labor, occupational health and safety, and the export of hazardous substances to developing countries. Publications of the agencies listed as resources in the preceding section of this Catalogue can be found in the original listing and are not included in the bibliography.

Barnet, Richard, and Ronald E. Muller. Global Reach: The Powers of Multinational Corporations. New York: Simon and Schuster, 1974.

Barovick, Richard L. "Labor Reacts to Multinationalism." Columbia Journal of World Business 5, no. 4 (July/August 1970).

Bernstein, Harold, and Joanne Bernstein. Industrial Democracy in 12 Nations. U.S. Department of Labor, Bureau of International Labor Affairs, Monograph No. 2. Washington, D.C., 1978.

Bosson, R., and B. Varon. The Mining Industry and the Developing Countries. New York: Oxford University Press, 1977.

Butler, Judy et al. "International Hazards: You Can Run But You Can't Hide." NACLA Report on the Americas, No. 12: Dying for Work. New York, March-April 1978, pp. 20-30.

Carmichael, Jack. "Interregional Symposium on Consideration of Environmental Quality in the Policy and Planning of Developing Countries." UNIDO, July/August 1977.

Castlemen, Barry. "The Export of Hazardous Factories to Developing Nations." International Journal of Health Services 9, no. 4 (1979):569-606.

_____. "How We Export Dangerous Industries." Business and Society Review, Fall 1978.

Center for Development Policy. "Campaign to Stop Runaway Shops." Preliminary Report No. 1, October 12, 1978.

Chase-Dunn, C. "The Effects of International Economic Dependence on Development and Inequality: A Cross-National Study." American Sociological Review 40 (1975):720-38.

Ciocca, Henry G. "Infant Formula Controversy: A Nestlé View." Journal of Contemporary Business 7, no. 4 (1978):37-56.

Conservation Foundation. A Conference on the Role of Environmental and Land-Use Regulation in Industrial Siting. Washington, D.C., June 21, 1979.

Curry, B. "U.S. Is Changing Emphasis in Aid to Developing Nations for Pesticides." Washington Post, May 14, 1977.

Curtis, Robert. Proposed Selective Tariff to Discourage Exportation of U.S. Health Hazards and Alleviate Negative Economic Impacts of Occupational and Environmental Health Standards. American Industrial Hygiene Conference, Salt Lake City, May 9, 1978.

Doyal, Lesley, and Imogen Pennell. The Political Economy of Health. London: Pluto Press, 1979.

Elling, Ray. Cross-National Study of Health Systems, Political Economies and Health Care. New Brunswick, N.J.: Transaction Books, 1979.

―――. "Industrialization and Occupational Health in Undeveloped Countries." International Journal of Health Services 7 (1977): 209-35.

Emara, A. M. "Occupational Health Problems in Agricultural Workers." Journal of Egyptian Medical Association 54 (1971): 314-21.

European Economic Community. Study of the Economic and Social Committee on Health and Environmental Hazards Arising from the Use of Asbestos. Brussels, February 22, 1979, pp. 28-33.

Evans, Peter. "National Autonomy and Economic Development: Critical Perspectives on Multinational Corporations in Poor Countries." International Organization 35, no. 3 (1971).

Flannery, Michael. "America's Sweatshops in the Sun." AFL-CIO American Federationist. Washington, D.C., 1979.

Forsyth, David J. C. "Restrictions on Multinationals in the Developing World." Multinational Business 4 (1977):1-7.

Frankel, Maurice. The Social Audit: Pollution Handbook. How to Assess Environmental and Workplace Pollution. London: Macmillan, 1978.

Galenson, Walter. Labour in Developing Economies. Berkeley: University of California Press, 1962.

Goldsmith, Frank. "The Significance of International Safety and Health Information: Its Role in the Establishment of Vinyl Chloride and Coke Oven Emissions Standards in the United States." Eighth World Congress on the Prevention of Occupational Accidents and Diseases, Bucharest, May 21-27, 1977.

Gould, William B. "The Rights of Wage Earners: Of Human Rights and International Labor Standards." Industrial Relations Law Journal 3, no. 3 (Fall 1979):489-516.

Grossman, Rachel. "Women's Place in the Integrated Circuit." Southeast Asia Chronicle 9, no. 66 (1978):3-17.

Gunter, Hans, ed. Transnational Industrial Relations. London: Macmillan, 1972.

Heller, Thomas. Poor Health, Rich Profits: Multinational Drug Companies and the Third World. London: Spokesman Books, 1977.

Hirayama, T. "Exporting Pollution." Kogai—The Newsletter from Polluted Japan, no. 2 (Winter 1974).

Instituto Centro Americano de Investigacion y Technologia Industrial. An Environmental and Economic Study of the Consequences of Pesticide Use in Central American Cotton Production. Project No. 1412, Guatemala, January 1977, p. 88.

Interamerican Development Bank. Multinational Investment, Public and Private, in the Economic Development and Integration of Latin America. Washington, D.C., 1968.

International Labor Office. The Relationship between Multinational Corporations and Social Policy. Geneva, 1972.

James, J. "Growth, Technology, and Environment in Less Developed Countries, A Survey." World and Development 6, no. 9 (July/August 1978):937-65.

Kenen, P. B., and R. Lubitz. International Economics. Englewood Cliffs, N.J.: Prentice-Hall, 1971.

Lall, S. "Medicines and Multinationals." Monthly Review 28 (March 1977):19-20.

Lean, Geoffrey. Rich World, Poor World. London: George Allen and Unwin, 1978.

Levenstein, Charles. "Political Economy of Occupational Safety and Health." New England Journal of Business and Economics 5 (1978):44-53.

Levinson, Charles. Capital, Inflation and the Multinationals. London: George Allen and Unwin, 1971.

Lim, Linda. "Women Workers in Multinational Corporations: The Case of the Electronics Industry in Malaysia and Singapore." Michigan Occasional Papers, no. 9 (Fall 1978).

London School of Hygiene and Tropical Medicine, T.U.C. Centenary Institute of Occupational Health. Proceedings of the Symposium on the Health Problems of Industrial Progress in Developing Countries. London, September 22-24, 1970.

Marsden, Keith. "Global Developing Strategies and the Poor." International Labor Review 117 (November/December 1978): 675-95.

Medawar, Charles. Insult or Injury? Marketing and Advertising of British Food and Drug Products in the Third World. London: Macmillan, 1979.

Mendeloff, John. Regulating Safety. Cambridge, Mass.: MIT Press, 1979.

Mulligan, J. E. "The Dominican Republic: Police State Protection for U.S. Corporations." WIN Magazine, no. 12 (February 2, 1976), pp. 8-12.

Navarro, Vincente. "The Economic and Political Determinants of Human (including Health) Rights." International Journal of Health Services 8 (1978):145-68.

_____. "The Underdevelopment of Health or the Health of Underdevelopment." International Journal of Health Services 4, no. 1 (1974):5-27.

Noweir, M. H. "Safe Level Criteria for Air Contaminants for Developing Countries." Bulletin of the High Institute of Public Health of Alexandria 5, no. 1 (1975).

Shaw, Robert d'A. "Foreign Investment and Global Labor." Columbia Journal of World Business 6, no. 4 (July/August 1971).

Silverman, M. The Drugging of the Americas. Berkeley: University of California Press, 1976.

Silvis-Cremer, G. K. "Asbestosis in South African Asbestos Miners." Environmental Research 3 (November 1970):310-19.

Solomon, Lewis D. Multinational Corporations and the Emerging World Order. Port Washington, N.Y.: Kennikat Press, 1978.

Stobaugh, Robert B. "The Multinational Corporation: Measuring the Consequences." Columbia Journal of World Business 6, no. 1 (January/February 1971).

Turner, Louis. Multinational Companies and the Third World. New York: Hill and Wang, 1973.

United Nations Industrial Development Organization. Industrial Development Priorities in Developing Countries: The Selection Process in Brazil, India, Mexico, Republic of Korea and Turkey. New York, 1979.

U.S. Congress. Report on Export of Products Banned by U.S. Regulatory Agencies. The Committee on Government Operations, Thirty-Eighth Report, Washington, D.C., September 27, 1978.

U.S. Department of Commerce. The Effects of Pollution Abatement on International Trade. Washington, D.C., April 19, 1975, pp. B-1 to B-52.

Vernon, Raymond. "Foreign Investors' Motivations in the LDC." Development Advisory Service: Economic Development Report: 172. Cambridge, Mass.: Harvard University Press, 1970.

Wionczek, Miguel S. Measures Strengthening the Negotiation Capacity of Governments in Their Relations with Transnational Corporations: Technology Transfer Through Transnational Corporations. New York: United Nations, 1979.

APPENDIXES

APPENDIX A: DATA COLLECTION METHODOLOGY

The following methods were used to facilitate data collection:

Conference contacts and networking: In September 1979 we constructed a questionnaire that was distributed to 200 people at the International Hazard Export Conference, November 1979. From October 1979 until March 1980 we attempted to identify and query the major public interest groups and international organizations through the following mechanisms:

> Personal contacts and references
>
> Mailing lists from academic, environmental, and occupational safety and health organizations and professional societies
>
> Information and data from professional journals, trade magazines, and government reports

Mail questionnaires: Follow-up questionnaires were (selectively) mailed to more than 300 international and national organizations and public interest groups.

Telephone surveys: Follow-up telephone surveys were made in place of or with the mail questionnaire to increase the response rate.

APPENDIX B: SAMPLE LETTER AND QUESTIONNAIRE

February 1, 1980

Please complete the Health and Safety Resource Catalogue Questionnaire printed on the reverse side of this page and return it to Jane Ives, 25 Elmore Street, Newton Centre, Massachusetts 02159 U.S.A. Include a listing of any publications, ongoing projects and/or programs. Please also list the names and addresses of other groups to contact for the catalogue.

We have undertaken this project in order to provide a catalogue of international health and safety resources for both workers and the occupational health community. This information will be published in the <u>International Occupational Safety and Health Resource Catalogue</u> for the National Institute for Occupational Safety and Health, Washington, D.C., U.S.A.

Your prompt return of this form will be greatly appreciated.

Thank you.

Jane Ives
25 Elmore Street
Newton Centre,
Massachusetts 02159
U.S.A.
(617) 964-7120

HEALTH AND SAFETY RESOURCE CATALOGUE: QUESTIONNAIRE

Please complete the following form and return it as soon as possible (by March 15) to Jane Ives, 25 Elmore Street, Newton, Massachusetts 02159, U.S.A. This information will be included

APPENDIX B / 191

and published in the <u>International Occupational Safety and Health Resource Catalogue</u>.

(where applicable)

Name_____ Organization_____

Work Address_____ _____

            _____ Contact_____

            _____ Position_____

Zip_____

Objectives and Goals of Organization:

Ongoing, Past, or Prospective Research/Projects:

Publications Available (including audiovisuals, periodicals):

References, names, and addresses of people/agencies working in occupational and environmental health and safety to contact, description of their work:

APPENDIX C: INTERNATIONAL CONFERENCE ON THE EXPORTATION OF HAZARDOUS INDUSTRIES, TECHNOLOGIES, AND PRODUCTS TO DEVELOPING COUNTRIES

An international conference on the Exportation of Hazardous Industries, Technologies, and Products to Developing Countries was held at Hunter College in New York City on November 2 and 3, 1979.

The conference, sponsored by the New Directions Program of the University of Connecticut Health Center and held in conjunction with the annual meeting of the American Public Health Association, focused on the public policy issues involved in the exportation of hazardous and polluting industries to nonregulating Third World nations to avoid the high costs of worker protection and environmental control in regulated, industrialized nations.

Among the international participants were Herman Rebhan, General Secretary, International Metalworkers Federation, Geneva, Switzerland; Dr. Gustavo Molina-Martinez, Colombia; Birger Viklund, Worklife Research Center, Stockholm, Sweden; and Zafrullah Chowdhury, Dacca, Bangladesh. These and other noted speakers and panelists at the two-day meeting examined regulatory processes in the United States, Europe, and developing countries and discussed specific case studies involving the exportation of asbestos mining and benzidine dye manufacturing, as well as the use of pesticides in Latin America.

The conference concluded with the adoption of five resolutions that support:

> the proposal of Dr. Anthony Robbins, Director of the National Institute for Occupational Safety and Health, to study occupational health hazards internationally and to report the findings to the International Labor Organization and other national and international organizations;
>
> the formation of an ongoing hazard export study group to examine the problems of the exportation of hazardous industries, products, and tech-

nologies to developing countries; runaway shops; and the reimportation of hazardous products;

the formation of an international conference of scientists, workers, union representatives, and other government and nongovernment organizations to study and discuss the effects of industrial hazards;

the right of people to know the generic identity and hazards of the substances with which they work, and the required labeling of substances in several languages;

the right of workers to have complete access to all their medical records as well as to health and safety test data in their own language;

the establishment of national health services to provide quality health care services to workers.

NEW DIRECTIONS PROGRAM OF THE UNIVERSITY OF CONNECTICUT HEALTH CENTER AND COSPONSORS OF THE CONFERENCE

American Labor Education Center, Washington, D.C.
Occupational Safety and Health Section and the International Health Section of the American Public Health Association, Washington, D.C.
Chicago Area Committee for Occupational Safety and Health, Chicago, Illinois
Department of Social and Preventive Medicine, Harvard Medical School, Boston, Massachusetts
Department of Environmental Health Sciences, Hunter College, New York, New York
Electronics Committee on Occupational Safety and Health, Mountain View, California
International Chemical Workers Union, Akron, Ohio
Massachusetts Coalition for Occupational Safety and Health, Boston, Massachusetts
Massachusetts Public Health Association, Boston, Massachusetts
Department of Social Medicine, Montefiore Hospital and Medical Center, Albert Einstein College of Medicine, Bronx, New York
National Council of Churches, New York, New York
National Institute for Occupational Safety and Health, Rockville, Maryland
New York Committee for Occupational Safety and Health, New York, New York

Philadelphia Area Project on Occupational Safety and Health, Philadelphia, Pennsylvania
Progressive Alliance, Washington, D.C.
United Auto Workers, Washington, D.C.
U.S. Environmental Protection Agency, Washington, D.C.
United Steelworkers of America, Washington, D.C.
Northeast Program of Cross National Studies of Health Systems, University of Connecticut Health Center, Framington, Connecticut
Urban Environment Conference, Washington, D.C.

PLANNING COMMITTEE

Jane Ives
Barry I. Castleman
Tina Borders
Stanley W. Eller

Ray H. Elling, Ph.D.
Dieter Koch-Weser, M.D.
Marshall S. Levine, M.D., M.P.H.
Charles Levenstein, Ph.D.

APPENDIX D: COMMITTEES ON OCCUPATIONAL SAFETY AND HEALTH (COSH GROUPS)

What is a COSH? A COSH is a nonprofit organization established by locals of many different unions in one geographic area. The board of directors is usually made up of representatives from the participating locals. Committees are set up to encourage the participation of doctors, lawyers, industrial hygienists, and other technical experts in the area.

How is a COSH financed? The basic support for a COSH office and small central staff generally is provided by regular dues from the participating local unions. Additional fees may be paid by each local union for special services, such as a training session. The dues and fees are usually very low because participating union members and professionals have other jobs and donate their time to the COSH. Most of the COSHes also have obtained small grants from foundations to pay for the cost of starting the COSH or for special projects. In addition, OSHA has given grants to a few COSHes under its "New Directions" program, which provides money to union groups and other organizations to improve their safety and health efforts.

What does a COSH do? A COSH generally does the following:

Sponsors training sessions on hazard recognition and control, federal and state laws, worker's compensation, and collective bargaining related to safety and health.
Researches potential hazards and suggests control methods.
Helps file and follow up complaints to OSHA and other federal and state agencies.
Provides assistance in drafting contract language on safety and health.
Refers medical, worker's compensation, and other problems to noncompany doctors and prolabor lawyers.

Organizes to demand strong enforcement of safety and health laws, stricter government standards, and improved worker's compensation programs.

Publicizes safety and health problems to unorganized workers and the general public through conferences and newspapers, radio, and TV.

How can my union and I participate in a COSH? If there is already a COSH in your area, contact its staff and they will tell you how you can get involved. If there is no COSH in your area, you should contact one of the COSHes on the list and ask them for advice. Ask them for suggestions about how to find other local unions and safety and health professionals in your area who might want to participate. Ask them for ideas on fund raising.

COSHes: How they fight job hazards. Local union leaders cite dozens of improvements in safety and health conditions as a result of cooperation between unions and their COSHes. A few examples are described below:

Teaching workers how to use the laws. Local 619 of the International Chemical Workers sent health and safety committee members to classes given by the Philadelphia-area COSH 9 (called PHILACOSH) on how to make effective use of OSHA. The unionists were taught the importance of keeping records of workers' health complaints, violations in the plant, and safety-related grievances that had been presented to the company. They were taught how to fill out the OSHA complaint form in a way that would get a quick response from OSHA. They were also told that they have a right to help make sure an OSHA inspector sees the important hazards in the workplace.

"Because we were specific in our complaint to OSHA, they sent somebody to inspect who knew something about our type of work," said Ernie Herbst, ICWU Local 619 health and safety committee member. "With what our COSH had taught us, we just took charge of the inspection, and made sure everything was getting seen. By the time it was over, OSHA had nailed the company with 92 violations on just about every major problem you can think of," he said. "They got more serious citations in some cases because our records proved that the company had known about some of the hazards but just didn't do much to clean them up."

Teaching about hazards. United Electrical Workers Local 242 sent members from a machine shop to one of the many training sessions conducted each year by the COSH in Massachusetts (MASSCOSH). The workers learned that exposure to benzene and to cutting oils containing nitrosamines may cause cancer. They presented the information to their employer and got the dangerous chemicals removed.

Researching hazards. Shop stewards from the International Ladies Garment Workers Union reported workers' complaints of back pains, stress, noise, and poor lighting to the New York COSH (NYCOSH). In many cases the stewards did not have enough technical knowledge to support their demands that the hazards be removed. NYCOSH provided industrial health experts who are working with the stewards to develop both evidence about the hazards and practical solutions to propose to employers.

Organizing to improve government enforcement. The COSHes have jointly mounted a campaign to strengthen workers' "right to know" about hazards they face. The campaign was started because many companies have failed to tell their workers about substances they are working with and to provide access to company safety and health records. The COSHes have organized demonstrations, coordinated testimony from many different local unions at OSHA hearings on the problem, and gathered thousands of signatures on petitions. Partly as a result of these efforts, OSHA has issued a new rule guaranteeing worker access to the injury and illness logs that all employees must keep. The COSHes also helped persuade OSHA to propose a new regulation requiring worker access to company medical records.

Organizing for safety: Other new approaches. In addition to setting up the COSHes, workers and unions are trying other new tactics for educating themselves on job safety and health and for putting more pressure on employers and the government. A few of those efforts are described below. Addresses and phone numbers are provided at the end of Appendix D so readers can contact the groups to learn more about their experiences.

Organize against one particular safety and health problem. An example is the Carolina Brown Lung Association, which was formed by active, disabled, and retired textile workers to combat the deadly lung disease caused by cotton dust. Like the COSH groups, the Brown Lung Association conducts medical screening, advises workers and local unions about their rights under safety and health laws, encourages workers to seek compensation for work-related disease, and brings nonunion workers in contact with union workers to cooperate on job safety and health problems.

Organize minorities or other special groups of workers whose problems cut across industry or union lines. Workers and their unions are bringing their questions about job-related health problems faced by women to the Women's Occupational Health Resource Center, which does research and helps conduct training sessions. The National Association of Farmworker Organizations is developing educational programs about safety problems, chemicals, and other hazards in agriculture. The National American Indian Safety Council is doing the same for Indian workers and tribal officials. SER/Jobs for Progress is developing training for Spanish-speaking workers in agriculture, construction, heavy metals manufacturing, and the service industries. Urban Environment Conference, a coalition of unionists, environmentalists, and urban activists, has helped labor groups conduct safety and health education for women and minorities.

Organize a school to train workers. Most universities still do not provide safety and health training for workers, and the programs that do exist are not always adequate. To provide their own training, a group of unions has started its own school called Pacific Northwest Labor College. Through the college, unions are able to pool the limited money they have for education, and to choose instructors and course material carefully. Many unions are getting help from other independent labor education projects. Most of these groups provide research on hazards, written and audio/visual educational materials, and workshops. The Highlander Center serves workers in Appalachia and the South. The Institute for Labor Education and Research in New York has two slide shows for rental, "Your Job or Your Life" and "Should Energy Cost an Arm and a Leg?" The Labor Safety and Health Institute also serves the New York City area. The Western Institute for Occupational/Environmental Sciences has a special project to help workers control exposure to asbestos. The Public Media Center produces materials on workers' safety and health rights and helps unions and public interest groups to conduct publicity campaigns. The Center for Occupational Hazards provides information about health hazards for workers in the arts.

Source: American Labor is published by the American Labor Education Center, a nonprofit institution set up by former union staff members with years of experience in providing training to workers. The Center can:

> help unions set up or improve programs on job safety and health, contract and labor law enforcement, membership education, public relations, or other subjects, and

produce materials for workers such as union publications and reports, slide/tape presentations, radio and newspaper ads, and posters.

If you or your union would like the Center's help, write to:

American Labor Education Center
1835 Kilbourne Place, N.W.
Washington, D.C. 20010

Or call (202) 462-8925 and ask for Matt Witt or Karen Ohmans. If you can get some people in your area interested, call a meeting and invite a representative from an existing COSH to come to speak and to answer questions.

If other unions can do it, so can yours. A COSH is an organization of unions and prolabor safety and health professionals. It usually includes locals from many different unions in one geographic area, as well as doctors, lawyers, scientists, journalists, organizers, and students. So far, COSHes have been set up by groups of locals in at least a dozen areas around the country.

Local unions that belong to a COSH may ask it to find professionals or experienced safety committee members from other unions who can help with specific problems on a volunteer basis. In addition, COSHes regularly hold training sessions both for individual locals and for workers from many locals who share common concerns. In all cases, the COSH's role is to provide workers and their unions with technical information and skills so they can solve safety and health problems themselves.

"It takes work at the local level, every day of the year, to accomplish anything in safety and health," said Mike Gaffney, health and safety committee member at United Auto Workers Local 6 and chairperson of the Chicago-area COSH. "The working people have to do it—you need a trained safety committee to find the problems and then keep after them," he said, "But the COSH gives us that training and backs us up when we run into technical things."

"Our COSH in the Philadelphia area gives us support from people who we can get hold of all the time," said Curt Wible, President of Oil, Chemical, and Atomic Workers Local 8-930. "Our international union, even though it is a leader in health and safety, couldn't possibly have the staff to provide everything every local needs—when we need it."

According to Marilyn Albert, safety committee member of the National Hospital Union, District 1199, the COSH in New York "serves a function no one union can serve. A COSH gives people from different unions a chance to meet and to learn from each other. Hazards like certain chemicals or radiation are the stock-in-trade of the health care industry, but they are also found in something as different as construction. There are solutions that have been found by one union that another union could use if they knew about it." . . . "Health and safety is also partly political—getting laws passed, defending them, and getting them enforced," she added. "A COSH gives trade unions a chance to work together on that too."

LIST OF COSH GROUPS

Chicago Area Committee on Occupational Safety and Health (CACOSH)
542 South Dearborn, No. 502
Chicago, Illinois 60605
(312) 939-2104

Electronics Committee on Occupational Safety and Health (ECOSH)
867 West Dana, No. 201
Mountain View, California 94041
(415) 969-8978 or 969-1545

Maryland Coalition for Occupational Safety and Health (MARYCOSH)
P.O. Box 3825
Baltimore, Maryland 21217

Massachusetts Coalition for Occupational Safety and Health (MASSCOSH)
P.O. Box 17326, Back Bay Station
Boston, Massachusetts 02116
(617) 482-4283

Minnesota Area Committee on Occupational Safety and Health (MACOSH)
1729 Nicollet Avenue, South
Minneapolis, Minnesota 55403
(612) 291-1815 (Tom O'Connell)

New Jersey Committee for Occupational Safety and Health (NJCOSH)
80 Central Avenue
Clark, New Jersey 07066
(201) 381-2469 (Mike McKowne)

New York Committee on Occupational Safety and Health (NYCOSH)
P.O. Box 3285, Grand Central Station
New York, New York 10017
(212) 577-0564 (Deborah Nagin)

North Carolina Occupational Safety and Health Project (NCOSH)
P.O. Box 2514
Durham, North Carolina 27705
(919) 286-2276

Philadelphia Area Project on Occupational Safety and Health (PhilaPOSH)
1321 Arch Street, No. 607
Philadelphia, Pennsylvania 19107
(215) 568-5188

Rhode Island Committee on Occupational Safety and Health (RICOSH)
P.O. Box 95, Annex Station
Providence, Rhode Island 02901
(401) 751-2015

SEACOSH (Seattle COSH)
3901 40th Avenue, SW
Seattle, Washington 98116
(206) 935-4497 (Vicki Laden)

Tennessee Committee on Occupational Safety and Health (TNCOSH)
Center for Health Services
Station 17
Vanderbilt Medical School
Nashville, Tennessee 37232
(615) 322-4773 (Jamie Cohen)

Western New York Council on Occupational Safety and Health (WNYCOSH)
59 Niagara Square Station
Buffalo, New York 14201
(716) 693-0165

Wisconsin Committee on Occupational Safety and Health (WISCOSH)
P.O. Box 92565
Milwaukee, Wisconsin 53202
(414) 962-2096

APPENDIX E: LABOR EDUCATION PROGRAMS IN THE UNITED STATES

University of Alabama at
  Birmingham
Center for Labor Education
  and Research
School of Business
University Station
Birmingham, Alabama 35294
(205) 934-2101

University of California,
  Berkeley
Labor Occupational Health
  Program
Institute of Industrial Relations
2521 Channing Way
Berkeley, California 94720
(415) 642-5507

University of California,
  Los Angeles
Center for Labor Research
  and Education
Institute of Industrial Relations
Los Angeles, California 90024
(213) 825-3537

University of Connecticut
Labor Education Center
Storrs, Connecticut 06268
(203) 486-3417

University of Hawaii
Center for Labor Education
  and Research
1420-A Lower Campus Road,
  Building 3
Honolulu, Hawaii 96822
(808) 948-7145

University of Illinois
Institute of Labor and Industrial
  Relations
504 East Armory
Champaign, Illinois 61820
(217) 333-0980

(Chicago Office)
Chicago Labor Education Program
1315 SEO Building
P.O. Box 4348
Chicago, Illinois 60680
(312) 996-2623

Indiana University
Division of Labor Studies
312 North Park
Bloomington, Indiana 47401
(812) 337-9082

University of Kentucky
Center for Labor Education
  and Research
643 Maxwelton Court
Lexington, Kentucky 40506
(606) 258-4811

University of Maine at Orono
Bureau of Labor Education
128 College Avenue
Orono, Maine 04473
(207) 581-7032

Antioch College
AFL-CIO Labor Studies Center
10000 New Hampshire Avenue
Silver Spring, Maryland 20903
(301) 431-6400

Michigan State University
Labor Program Service
School of Labor and Industrial
  Relations
South Kedzie Hall
East Lansing, Michigan 48824
(517) 344-5070 or 355-2214

University of Minnesota
Labor Education Service, IRC
447 BA Tower
271-19th Avenue South
Minneapolis, Minnesota 55455
(612) 373-3662, -4110, -5380, -5306

Rutgers University, The State
  University of New Jersey
Labor Education Center
Institute of Management and
  Labor Relations
University Extension Division
Ryders Lane and Clifton
  Avenue
New Brunswick, New Jersey
  08903
(201) 932-9502

Cornell University
New York State School of Industrial and Labor Relations
Division of Extension and
  Public Service
Ithaca, New York 14850
(607) 256-3281

(Metropolitan District Staff)
New York State School of Industrial and Labor Relations
7 East 43rd Street
New York, New York 10017
(212) 697-2247

(Western District Staff)
New York State School of Industrial and Labor Relations
120 Delaware Avenue, Room 225
Buffalo, New York 14202
(716) 842-4270

Ohio State University
Labor Education and Research
  Service
1810 College Road
Columbus, Ohio 43210
(614) 422-8157

University of Oregon
Labor Education and Research
  Center
Eugene, Oregon 97403
(503) 686-5054

University of Wisconsin,
  Extension
School for Workers
One South Park Street, No. 701
Madison, Wisconsin 53706
(608) 262-2111

APPENDIX F: EDUCATIONAL RESOURCES CENTERS IN THE UNITED STATES

The National Institute for Occupational Safety and Health has funded 12 regional Educational Resources Centers (ERCs) aimed at providing graduate and undergraduate degree programs and continuing education programs through short courses and workshops, in the disciplines of occupational safety, industrial hygiene, occupational medicine, and occupational health nursing. In addition, programs of outreach are maintained to provide assistance to other academic and professional institutions.

ERCs are contributing to the prevention of occupational injury and disease through programs aimed at providing professional development, professional manpower, and technical and scientific assistance in occupational safety and health for labor, management, and government.

Arizona Educational Resource Center
University of Arizona
Arizona Health Sciences Center
1145 North Warren (ACOSH)
Tucson, Arizona 85724
(602) 626-6835
Herbert K. Abrams, M.D., Director

California Educational Resource Center
University of California, Irvine
Department of Community and Environmental Medicine
Irvine, California 91717
(714) 833-6269
B. Dwight Culver, M.D., Director

Cincinnati Educational Resource Center
University of Cincinnati
Institute of Environmental Health
3223 Eden Avenue
Cincinnati, Ohio 45267
(513) 872-5701
Raymond R. Suskind, M.D., Director

Harvard Educational Resource Center
Department of Environmental Health Sciences
Harvard School of Public Health
665 Huntington Avenue
Boston, Massachusetts 02115
(617) 732-1260
David Wegman, M.D., Director

Illinois Educational Resource Center
University of Illinois
School of Public Health
P.O. Box 6998
Chicago, Illinois 60680
(312) 996-2591
Bertram W. Carnow, M.D., Director

Johns Hopkins Educational Resource Center
Johns Hopkins University
School of Hygiene and Public Health
615 North Wolfe Street
Baltimore, Maryland 21205
(301) 955-3900 or 3720
Gareth M. Green, M.D., Director

Minnesota/Iowa Educational Resource Center
University of Minnesota
School of Public Health
420 Delaware Street, S.E.
Minneapolis, Minnesota 55455
(612) 373-8080
Conrad P. Straub, Ph.D., Director

New York/New Jersey Educational Resource Center
Mt. Sinai School of Medicine
1 Gustave Levy Place
New York, New York 10029
(212) 650-6174
Irving J. Selikoff, M.D., Director

North Carolina Educational Resource Center
University of North Carolina
School of Public Health
Chapel Hill, North Carolina 27514
(919) 966-1023
David A. Fraser, Sc.D., Director

Texas Educational Resource Center
The University of Texas Health Science Center at Houston
School of Public Health
P.O. Box 20186
Houston, Texas 77025
(713) 792-4312
Marcus M. Key, M.D., Director

Utah Educational Resource Center
Rocky Mountain Center for Occupational and Environmental Health
University of Utah Medical Center
DFCM Room BC 106
Salt Lake City, Utah 84132
(801) 581-8719
William N. Rom, M.D., Director

Washington Educational Resource Center
University of Washington
Department of Environmental Health SC-34
Seattle, Washington 98185
(206) 543-6991
John T. Wilson, Jr., M.D., Director

APPENDIX G: OSHA NEW DIRECTIONS PROGRAMS

I am pleased to have the opportunity today to address the important compliance assistance aspects of our federal occupational safety and health program—namely, education, consultation, and informational assistance—and on Friday to continue these hearings with a discussion of our enforcement efforts.

The problems in occupational safety and health are among the most complex and difficult our society faces. They involve interactions among numerous factors, including:

> the technology, materials, and processes of the workplace;
> the law as it evolves through administrative initiative and court decision; and
> human beings in their various roles as employers, first-line supervisors, workers, and union leaders.

Problems such as these do not yield to single or simplistic approaches; they require an orchestrated array of policy and programmatic instruments playing in concert. Indeed, an orchestra is an apt analogy here. Just as the string section is the essential centerpiece of a symphony orchestra, so is enforcement, based upon mandatory standards and effective inspections, the foundation of a national workplace safety and health program. Just as strings alone are not enough for an orchestra, so OSHA also needs its consultative services, its education and training program, its public information initiatives, and the informal day-to-day advice and assistance offered throughout the country to employers and employees by its professional staff if it is to be truly effective in reducing workplace

---

Statement of Basil Whiting, Deputy Assistant Secretary of Labor for Occupational Safety and Health, before the Committee on Labor and Human Resources, U.S. Senate, March 18, 1980.

illnesses and injuries and in achieving a broad base of public awareness and acceptance.

While we recognize this today, earlier administrations I must admit did not. There was no federal support of consultative services as such until FY [fiscal year] 1976. Until FY 1977, no funds were specifically designated in the OSHA budget for employer and worker training. While some funds in FY 1976 were actually expended on training, they amounted to only 0.07 percent of OSHA's funds, or $1 million. In the proposed FY 1981 budget, OSHA's commitment to those essential services will have increased by a factor of fourteen—to $13.9 million. For the onsite consultation program, the commitment of funds will have more than doubled over that same period, from $9 to $23 million. A total of $40 million will be directed toward all compliance assistance activities in FY 1981.

In short, this may be the tenth year of OSHA's existence, but as far as consultation and education are concerned, OSHA can be considered an agency that is only three years old.

COMPLIANCE ASSISTANCE

Section 21(c) of the Occupational Safety and Health Act assigns to the secretary of labor responsibility for the "education and training of employers and employees in the recognition, avoidance, and prevention of unsafe or unhealthful working conditions," as well as the authority to "consult with and advise employers and employees, and organizations representing employers and employees as to effective means of preventing occupational injuries and illnesses."

With 2,580 federal and state safety health inspectors to protect 4.4 million workplaces that employ 65 million workers, it is only ordinary good sense that to achieve its statutory objective, OSHA must enlist the active interest and cooperation of all concerned.

INFORMAL ASSISTANCE

Before we go on to discuss the formal programs in which you have expressed an interest, I would like to describe for you some of the day-to-day assistance that OSHA provides to employers and workers throughout the country.

At present, OSHA area offices in all jurisdictions receive hundreds of telephone inquiries every day from employers and workers concerned about workplace safety and health. Frequently employers will stop by in person to discuss ways of meeting complex requirements for hazard abatement. Each regional office has technical support personnel to assist in solving problems large and small. For example, several regions have noise experts who can advise on the latest methods to reduce noise. For those that do not, OSHA makes experts available on an as-needed basis. Such experts spend a large part of their day assisting with workplace noise problems. Other technical staff devote much of their time to specific health and safety problems of employers and employees.

Through telephone calls, meetings, and seminars, OSHA's technical support and compliance staff in the area, regional, and national offices advise daily on abatement of hazards, interpretation of standards, and possible variances to those standards. For example, inquiries about protective guarding around machines such as punch presses and press brakes are frequent. Our records show that some 200,000 hours of our field compliance personnel time were spent in this sort of informal compliance assistance.

OSHA periodically alerts employers and workers to potentially harmful substances or hazardous workplace situations that have only recently come to light. For example, when OSHA found an unusual incidence of sickness among workers in a plant in which ESN (Dimethylominopropion Bis[2-(dimethylomino)ethyl] ether) was used as a catalyst in the manufacture of electrical equipment, the agency immediately sent telegrams to manufacturers of similar products alerting them to the danger. ESN is known to cause kidney and bladder damage. Since then, a reasonably safe substitute for ESN has been found, and workers are no longer subjected to this particular hazard.

For hazards in grain elevators for which no single configuration of regulations exists, OSHA has prepared and distributed 15,000 pamphlets telling how to reduce the threat of explosion and of overexposure to pesticides. Among the dangers employers and workers in these facilities are told to look for are faulty electrical wiring and excessive concentrations of dust. OSHA worked closely in this effort with those officials in the Environmental Protection Agency and the Department of Agriculture, particularly in that agency's Federal Grain Inspection Service, who are most knowledgeable about this industry.

In the past few months OSHA and the National Institute for Occupational Safety and Health (NIOSH) have issued an intelligence bulletin alerting employers and workers to the hazards of excessive exposure to low-level radio frequency radiation.

Workers using heatsealers to wrap meat and produce in supermarkets, for example, may be exposed to this risk.

In addition to these specific hazard alerts, OSHA and NIOSH, singly and jointly, have published hundreds of similar pamphlets, as well as booklets and other informational materials designed to help employers and workers understand their rights and responsibilities and eliminate and control workplace hazards.

It is OSHA's present policy to publish two versions of each brochure dealing with specific safety and health topics, one directed to employers, the other to workers. In addition, we have published a number of booklets aimed separately at each of these audiences, such as those informing workers of their rights under the act, or our highly successful Small Business Handbook, which is aimed at helping owners and managers of small firms meet their obligations under the act. These handbooks have been in demand by members of Congress for their constituents' use, as well as in the small business community at large. More than 2 million copies have been distributed to date.

"NEW DIRECTIONS"

History of the Program

Before the current administration [that of President Carter] assumed office, OSHA had established only a limited number of education and training projects under contract with nonprofit organizations to train certain workers and employers in high-hazard industries in workplace health and safety. Those projects suffered from the absence of a coordinated OSHA education and training program, as well as from a lack of the long-term federal commitment necessary to make the training effort an integral and self-sustaining part of each organization's operations. Nonetheless, the projects provided a valuable lesson: They made clear to OSHA that the most practical way to get employers and workers to recognize and control workplace hazards was not to provide training directly, but to help key organizations develop the capability not only to provide the training themselves, but to build into their organizational structures the competence needed to deal effectively with problems of health and safety in the workplace.

In April 1978, OSHA launched a major effort to develop this needed institutional competence through its "New Directions" grants program. The objective of the new program is to utilize pivotal organizations to provide job safety and health education and related services to workers and employers. Labor unions, trade associations, educational institutions, and nonprofit organizations are eligible to apply for these grants.

Included in the broad range of activities such groups may carry out with OSHA funds are:

> training in employer and worker rights and responsibilities, and hazard recognition and control,
> technical assistance in hazard identification and abatement, and
> development and dissemination of informational materials.

Two kinds of grants are offered: planning grants for an initial year, and developmental grants for a duration of up to five years. Grants are awarded competitively. Special attention is given to recruiting applications from organizations particularly suited to serve the needs of small businesses, of workers in hazardous occupations, and of minorities. Various experts drawn from labor unions, trade associations, and universities screen new applications and make recommendations for awards on the basis of established criteria.

The Present Program

Within six months of the program's initiation, 86 New Directions grants totaling $5.8 million were awarded. In 1980, most of these were refunded and a second round of grants for new applicants is planned shortly. Funds granted for the program this year total $11 million. The expectation is that within a few years, most of these projects should be able to operate on a self-sustaining basis without need for further federal assistance. Recipient organizations are using their grants to develop educational materials, conduct training sessions, and design strategies to identify and resolve specific workplace safety and health problems. These are groups that only 18 months ago had, in general, virtually no competence to deal with workplace safety and health problems, but now, as a direct result of the grants program, are able to include 280 safety and health professionals on their staffs. During the first year of operation alone, over 20,000 employers and workers received intensive instruction in job safety and health; countless more were exposed to training sessions, conferences, and informational materials.

This year more than 230 applications have been received for the proposed second round of grants. Our advisers are evaluating these applications, and awards for the coming year should be announced within the next few months.

Quality of the Program

OSHA has begun to assess the effectiveness of current New Directions projects by reviewing grantees' quarterly reports and by visiting projects to observe them in action. At the same time, the agency is designing a comprehensive, long-range evaluation plan. That plan may involve a sampling of the program's impact in affected workplaces as well as a thorough survey of institutional changes in the recipients that might lead to safer and healthier workplaces. Changes we might look for include:

> increased numbers of professional health and safety personnel,
> new technical services addressing specific health and safety problems, and
> integration of occupational safety and health issues in ongoing organizational activities such as collective bargaining, association newsletters, or training offered to key leaders.

Our immediate purpose is to fill the need in recipient organizations for proficiency in occupational safety and health—to develop their capacity, through education and training and related services, to deal effectively on a self-sustaining basis with safety and health problems in the workplace. After only 18 months of the program, initial progress reports from the grantees are heartening. There are encouraging signs that the capabilities we are seeking to foster in these organizations are in fact developing. Let me give you a few instances of successful results experienced thus far.

New Directions "Success Stories"

Case 1: Pennsylvania Foundrymen's Association

I understand that Bruce Eckert of the Pennsylvania Foundrymen's Association will be appearing before you later this morning. They have done a fine job with their New Directions grant, and I hope I won't be stealing his thunder if I tell you a little bit about what they have accomplished. The association conducted a series of health and safety seminars covering problems identified in a survey of nine member foundries, each of which employs from 100 to 500 workers. High-level management from those firms attended the seminars.

As a direct result of the seminars, significant steps were taken to improve health and safety conditions. For example, one of the foundries initiated formal accident investigation procedures as well as regular audiometric testing for all workers. Another foundry appointed a safety director for the first time and hired an outside consultant to improve the ventilation system. Still another established the practice of periodic safety meetings and took steps to provide medical examinations for all workers, both before hiring and during employment at the plant. It is fair to say that these changes were prompted by the training and follow-up services provided by the safety and health staff of the Pennsylvania Foundrymen's Association.

Case 2: Indiana University

Indiana University provides an example in which formal instruction under the New Directions program led to specific improvements in workplace safety and health. At a class conducted by a New Directions grantee under the auspices of the university, employees of a large electronics manufacturing company were given information on safety procedures, including the handling of hazardous substances. In their plant it was the practice for workers to rock large carboys in which sulphuric acid was stored in order to pour the acid. In the process, acid frequently spilled and a number of burns had occurred. From what the workers learned at the class about the transfer of hazardous substances, they were able to suggest to management that vacuum pumps be used to transfer the acid. All the carboys are now equipped with such pumps, and acid spills and burns are no longer regular occurrences.

Case 3: International Association of Machinists and Aerospace Workers

Last summer a business representative of the machinists union investigated a situation in which 15 electricians complained of dizziness and nausea on the job. Since a chemical cause of the sickness was suspected, the union representative sent the label of a coolant used in the transformers on which the electricians worked to the safety and health department of the union, which has a New Directions grant. There it was discovered that the coolant was PCB, a toxic carcinogenic chemical. With this information, labor and management were able to deal jointly with the hazard. A different and safer coolant was substituted, and the problem was solved.

ON-SITE CONSULTATION

Many employers need help in developing a better understanding of their obligations under federal and state occupational safety and health acts. The ability to recognize hazards and to interpret complex standards is particularly difficult for small employers, many of whom are already overburdened with the work involved in running their own businesses and who may lack resources to obtain the services of industrial hygienists and safety engineers. While informal assistance of the kind I described earlier is invaluable, as are the educational and training opportunities available to trade associations under the New Directions program, there is also a clear need for formal assistance on an individual worksite basis. This need is one that the Congress recognized and that OSHA first began to address in 1975. The onsite consultation program launched that year now provides advice on workplace safety and health to employers throughout the country, upon request and at no charge to them.

History of the Program

For the first four years of OSHA's existence, onsite consultative services were offered only by those states with state-administered, federally approved occupational safety and health programs (state-plan states) at the normal 50 percent funding rate. In response to a demand for similar services in states under the federal occupational safety and health program (federal OSHA states), regulations were issued in May 1975 that made possible 50 percent federal funding for consultation projects in those federal OSHA states that wished to provide such services through state personnel. Federal OSHA provided these funds by entering into agreements with states under section 7(c)(1) of the act; hence this onsite consultation program is generally referred to as the "7(c)(1) program."

By the end of FY 1976, 22 state-plan states and 12 federal OSHA states were providing consultative services. In the first six months of the federal 7(c)(1) program (from January through June of 1976), 151 consultants made 3,417 onsite visits in states with federal projects in response to 4,902 requests from employers.

At the suggestion of the Senate conferees on the Labor-HEW Appropriations for FY 1977, the agency increased the federal contribution in the 7(c)(1) agreements to 90 percent in order to encourage more states to participate.

As a result, last year, consultative services were available, upon request and free of charge, to employers throughout the country as well as in Puerto Rico and the Virgin Islands In 1980 onsite consultative services will be extended to the Pacific Islands area.

In those few states and jurisdictions that have chosen not to participate in the program, onsite consultative services are provided by direct, 100 percent federally funded, private contracts. Table G.1 shows the present participation in the onsite consultation program by state and indicates the nature of the funding provided.

State-plan states that agree to observe the requirements of the 7(c)(1) program are now authorized to receive 90 percent funding. In the coming year OSHA will be working to achieve greater uniformity in the program so that comparable consultative services may be offered to employers in all parts of the country.

The program has grown rapidly over the past three years. As word spreads among employers, the demand for these services grows. Last year, one out of every six OSHA-related visits to a worksite was a consultative visit and not an inspection. OSHA regulations require that in scheduling consultative visits, priority be given to small business. More than half—56 percent—of such visits were made in workplaces with 20 or fewer employees.

In 38 jurisdictions with federal projects, 18,000 requests for onsite consultation were received; 20,865 such visits were made in FY 1979, clearing up a backlog of requests from the previous year. By the start of FY 1980, the backlog had been reduced to only 585. More than 300 consultants, including 110 industrial hygienists, conducted these visits. In addition, 17,422 consultations were made in the remaining 16 state-plan jurisdictions, which had their own consultation programs. (It must be noted that a "consultation" is defined differently for reporting purposes among these states and therefore this total cannot be compared to the federal 7(c)(1) tally.)

Tables giving further details of OSHA's onsite consultation program are included at the end of the text.

What Onsite Consultation Involves

Let me take a moment to explain what a consultative visit to a worksite involves. Such visits are made only at the request of employers and at no cost to them. The purpose of the visits is to help employers identify and correct hazardous workplace conditions without fear of penalties. The consultant is trained under OSHA guidelines so that the advice to employers will not be contradicted by an OSHA inspector.

The consultant prepares himself for the visit so that he is conversant with the processes and materials used by the firm. The visit begins with an opening conference with the employer in which the consultant explains the scope of the program, reminds the employer of his obligations under the act, and explains that if he discovers an imminent danger, it must be corrected immediately and that the employer is expected to eliminate any serious hazards within a reasonable abatement period. Employers failing to take the necessary actions to correct serious problems are referred to OSHA for possible enforcement action.

In the scores of thousands of OSHA-funded onsite consultative visits, there have been only 12 instances of such referrals. The consultation program is not enforcement-oriented, but, consistent with the intentions of the act, it does oblige employer action if serious hazards threatening workers' lives are found.

The consultant encourages the employer to allow full employee participation in the next stage of the visit—a walkthrough of the company facility, but the extent of such participation again rests on the employer's consent. As they walk through the workplace, the consultant identifies hazards on the spot and may make suggestions for their elimination, although such suggestions are more likely to occur in the closing conference with the employer.

In the closing conference, if hazards are found, their elimination or control is discussed. The consultant forwards a written report to the employer, usually within 15 working days. The report includes all findings, suggested remedies, and an agreed-upon deadline for correction of each serious hazard found.

The program is designed so that information obtained through the visit will remain confidential and will not be used to trigger an OSHA inspection. Since the consultation is being conducted at the employer's request, his consent is necessary at every stage of the visit.

Publicizing the Program

States have an obligation under the program to encourage requests from employers by publicizing the availability of consultative services through such means as paid newspaper advertisements, articles in trade journals, direct mailings, telephone solicitations, and public service announcements on radio and television. Promotional efforts are being directed particularly toward smaller businesses in high-hazard industries.

Quality of Performance

OSHA has encouraged technical training of consultants in the compliance procedures and policies of the agency. Many consultants take the introductory training for compliance safety and health officers at the Occupational Safety and Health Institute. All project consultants and their supervisors are trained at the University of Alabama in the proper classification of hazards, in program-operating guidelines, and in the communicating skills necessary for dealing effectively with business. This year, additional courses are being offered at that university to extend safety consultants' ability to address occupational health problems. This "cross-training" for safety consultants includes courses in health-hazard identification, abatement, and control methods and is similar to the cross-training provided to compliance safety officers. Another course provided this year to health consultants should result in increased use of laboratory samples of suspected workplace contaminants.

The performance of each state consultation project is monitored at present by OSHA's regional staff, who conduct reviews of project offices four times a year to assess operations and examine case files. In addition, most projects have self-monitoring procedures established under OSHA guidelines. Twice a year, special project personnel accompany consultants on site and prepare evaluations that are reviewed by the OSHA regional office. With the permission of the employer, these special project personnel may also make spot-check follow-up visits to sites previously visited by a consultant in order to assess the caliber of the advice given to the employer. In some projects having too few consultants to justify a special monitor, a private contractor is used.

Consultation "Success Stories"

Perhaps the best way I can tell you of the success of the onsite consultation program is to cite some specific examples from the state project files:

Case 1: Alabama

Among the hundreds of commendatory letters received by the Alabama Consultation Project, there are tributes like these:

"Your organization saved the hosiery mills in this area roughly $15,000. . . ."

"I feel more programs of this caliber are necessary for industry and could go a long way in reducing injuries and accidents."

"I feel this is one of the most helpful government programs I have had any experience with and would recommend your service to any industry."

Here is an account of the Alabama Consultation Project in action. On a Friday evening, ten workers on the second shift in the quality control testing room of a magnetic tape manufacturing plant suddenly began to complain of headaches, dizziness, blurred vision, and flushed skin. Two workers fainted and were taken to the local hospital. The entire plant, which normally operated around the clock, seven days a week, was shut down. Plant engineers, safety staff, and nursing personnel all were mystified since chemicals were not used in the workroom, and there had been no illness in other areas of the plant.

An industrial hygienist from the Alabama Consultation Project arrived at the remote, rural plant site within three hours. He took air samples for various contaminants; all tests proved negative. He then suggested that carboxyhemoglobin tests be performed on all affected workers. The tests established that there were elevated levels of this substance in the workers' blood, and carbon monoxide was identified as the probable cause of their illness.

By midnight, the consultant had discovered the existence of a previously unknown passageway that connected the quality control room to another room in a distant area of the plant. Major construction activity had been under way in the other room at the time that the workers in the quality control room became ill. The equipment used in the construction activity gave off large amounts of carbon monoxide, which found its way through the abandoned passageway to the quality control room.

The plant supervisor saw that air flow was blocked through the passageway, and work resumed on the third shift with no further incidents.

Case 2: Colorado

The consultation project at Colorado State University reports that in the first six months of its program, from June 1979 to January 1980, four safety and health consultants conducted 132 onsite visits affecting 5,072 employees. They identified 178 serious hazards, ranging from overexposure to toxic chemicals to faulty electric wiring—hazards that could have led to debilitating disease, death, or permanent disability. All the hazards thus identified have either been eliminated or are in the process of being corrected.

As an example of their success, they refer to a health survey performed by one of their consultants at a screen printing shop in Colorado. This survey revealed that employees were exposed to potentially harmful levels of methyl ethyl ketone (MEK) vapors during the cleaning of the printing screens. MEK has been associated with the development of peripheral nerve damage in workers in certain factories using it as a solvent. At the suggestion of the industrial hygiene consultant, the company discontinued the use of MEK and is now using a bleach-soak process to clean the screens. The bleach in use does not generate air contaminants and cleans the screens as effectively as the more toxic MEK.

Case 3: Utah

A Utah state consultant phoned a businessman about the consultation program because of the possible lead hazards in his industry. The employer expressed interest, saying that he had had consultative advice previously from an insurance company representative regarding lead-exposure problems, but that the advice had not been specific enough for him to develop suitable remedies.

The visit was agreed upon, and an industrial hygienist from the Utah Consultation Project visited the plant. There he took air samples that revealed that workers were exposed to excessive levels of airborne lead as well as of noise. For the lead exposure, the consultant recommended installation of exhaust ventilation along with a rigidly enforced respirator program, other personal protective equipment, and medical surveillance. At the consultant's advice, the entire workforce received special training in hearing conservation, and assignments to noisy areas were restricted so as to limit employee exposure. At the same time, use of protective hearing devices was made mandatory. Whereas before the consultant's visit workers resisted using such devices, the special training convinced them of the need to protect their hearing, and they willingly accepted the devices. In fact, even employees not working in high-noise areas requested them.

CONCLUSION

In conclusion, then, we might say that it was not until the seventh year of OSHA's existence as an agency that an appropriate array of policies and program instruments finally began to come into play.

Consultation and education are essential elements of a successful national occupational safety and health program because:

Consultation and training programs assist and support employers and employees in fulfilling their responsibilities to provide healthier and safer working conditions. Only the constant vigilance of enlightened employers and an enlightened workforce, aware of the hazards in the workplace and committed to reducing workplace injuries and disease, can assure the safe and healthful working conditions that are every worker's right.

These programs provide a vital service for the small businesses that make up the vast majority of the nation's workplaces. Owners and managers of small firms often lack the financial resources to obtain advice from industrial hygienists and safety engineers. Recognizing hazards in the workplace and interpreting complex health standards can pose difficulties for all employers, but the problem is particularly troublesome for small business employers.

Thank you. I would be happy to answer any questions you may have.

TABLE G.1

OSHA-Funded Consultation Projects, March 1980

Sec. 7(c)(1) Program (90/10 Funding)	State-Plan Programs (50/50 Funding)	Direct Contracts with Private Firms (100 Percent Funding)
Alabama	Arizona	Idaho
Alaska[a]	Hawaii	Louisiana
Arkansas	Indiana	New Hampshire
California[a]	Kentucky	Pennsylvania
Colorado	Maryland	South Dakota
Connecticut[b]	Michigan	American Samoa[c]
Delaware	Minnesota	Guam
Florida	Nevada	Trust Territories
Georgia	New Mexico	
Illinois	North Carolina	
Iowa[a]	Tennessee	
Kansas	Washington	
Maine	Wyoming	
Massachusetts	Puerto Rico	
Mississippi	Virgin Islands	
Missouri		
Montana		
Nebraska		
New Jersey		
New York		
North Dakota		
Ohio		
Oklahoma		
Oregon[a]		
Rhode Island		
South Carolina[a]		
Texas		
Utah[a]		
Vermont[a]		
Virginia[a]		
West Virginia		
Wisconsin		
District of Columbia		

[a]State-plan states.

[b]Connecticut has a 7(c)(1) consultation program for privator-sector employees, an 18(b) (state-plan) program for public-sector employees.

[c]Will be announced for bids during FY 1980.

TABLE G.2

Onsite Consultation Visit Summary, Federal 7(c)(1) Program, Number of Visits by State Project

	FY 1979	FY 1978	FY 1977
7(c)(1) Agreements with Nonplan States			
Alabama	776	289	—
Arkansas	678	1,120	598
Colorado	79	—	—
Connecticut	907	476	—
Delaware	303	220	259
District of Columbia	279	363	127
Florida	1,442	—	—
Georgia	141	—	—
Illinois	945	289	—
Kansas	547	468	594
Maine	111	—	—
Massachusetts	998	883	1,680
Mississippi	49	—	—
Montana	141	—	—
Nebraska	287	284	300
New Jersey	439	244	—
New York	2,318	3,268	3,575
North Dakota	3	—	—
Ohio	1,350	1,375	794
Oklahoma	1,115	1,409	622
Rhode Island	47	—	—
Texas	1,073	1,232	598
West Virginia	965	824	1,576
Wisconsin	596	571	603
7(c)(1) Agreements with 18(b) Plan States			
Alaska	177	251	—
California	3,334	2,567	—
Iowa	19	—	—
Oregon	416	162	—
South Carolina	588	370	—
Utah	194	—	—
Vermont	0	—	—
Virginia	275	217	—
Contractor projects	273	—	—
Total visits	20,865	16,882	11,326

(Rev. 2/14/80)
OPLIP, 3/15/80

TABLE G.3

Onsite Consultation Visit, FY 1979, Federal 7(c)(1) Program
(percent by business size)

	Number of Workers				
State	1-10	11-25	26-49	50-99	100+
7(c)(1) Agreements with Nonplan States					
Alabama	23	27	21	11	18
Arkansas	27	22	26	15	17
Colorado	23	37	12	23	5
Connecticut	38	23	20	9	10
Delaware	46	23	13	9	10
District of Columbia	49	18	17	7	9
Florida	28	34	20	10	8
Georgia	2	7	11	16	64
Illinois	25	25	20	15	14
Kansas	34	30	21	10	5
Maine	3	5	22	60	10
Massachusetts	20	30	10	20	20
Mississippi	28	0	7	18	47
Montana	52	28	14	6	0
Nebraska	59	25	8	5	3
New Jersey	29	23	23	12	13
New York	25	47	11	11	6
North Dakota	34	0	0	33	33
Ohio	34	33	13	10	10
Oklahoma	44	22	14	9	11
Rhode Island	24	20	20	14	22
Texas	21	21	26	15	17
West Virginia	34	39	15	7	5
Wisconsin	24	17	16	13	30
7(c)(1) Agreements with 18(b) Plan States					
Alaska	52	20	13	13	2
California	30	20	21	15	14
Iowa	54	15	15	1	15
Oregon	47	10	8	10	25
South Carolina	20	21	13	13	33
Utah	39	27	9	7	18
Vermont	52	22	22	1	1
Virginia	21	22	16	13	28
Nationwide average	33	22	16	13	16

Percentage of Visits by Size of Business

Category	Percent	Cumulative
10 or fewer workers	36	36
11 to 20 workers	20	56
21 to 50 workers	22	78
51 to 100 workers	9	87
over 100 workers	13	100

Office of Consultation Programs
March 4, 1980
OPLIP, March 15, 1980

TABLE G.4

Onsite Consultation Visits, FY 1979, in 18(b) States under 23(g) Grants

Arizona	380	Nevada	1,037
Connecticut	166*	New Mexico	136
Hawaii	247	North Carolina	1,670
Indiana	165	Puerto Rico	44
Kentucky	234	Tennessee	218
Maryland	429	Virgin Islands	1
Michigan	11,420	Washington	880
Minnesota	304	Wyoming	91
		Total	17,422

*In workplaces of public-sector employees.

Office of State Programs
January 3, 1980

APPENDIX H: ORGANIZATIONAL LISTING OF NATIONAL AND INTERNATIONAL UNION CONTACTS, SELECTED AFL-CIO DEPARTMENTS, AND FEDERAL ADVISORY COUNCIL ON OCCUPATIONAL SAFETY AND HEALTH

Actors and Artists of America, Associated
Frederick O'Neal, President
1500 Broadway
New York, New York 10036
(212) 869-0358

AFL-CIO (see separate page of Selected Departments)
815 16th Street, N.W.
Washington, D.C. 20006
(202) 637-5000

Air Line Pilots Association
John J. O'Donnell, President
1625 Massachusetts Avenue, N.W.
Washington, D.C. 20036
(202) 797-4000

Aluminum Workers International Union
L. A. Holley, President
Suite 711, Paul Brown Building
818 Olive Street
St. Louis, Missouri 63101
(314) 621-7292

Asbestos Workers, International Association of Heat and Frost Insulators and
Roy Steinfurth
505 Machinists Building
1300 Connecticut Avenue, N.W.
Washington, D.C. 20036
(202) 785-2388

Automobile, Aerospace and Agricultural Implement Workers of America, United
Frank Mirer
8000 E. Jefferson Avenue
Detroit, Michigan 48214
(313) 926-5321

Bakery, Confectionery and Tobacco Workers International Union
Vaughn Ball
1828 L Street, N.W.
Suite 900
Washington, D.C. 20036
(202) 466-2500

Boilermakers, Iron Ship Builders, Blacksmiths, Forgers, and Helpers, International

---

U.S. Department of Labor, Occupational Safety and Health Administration, Washington, D.C., January 12, 1980.

Brotherhood of
Michael Wood
New Brotherhood Building
8th Street, at State Avenue
Kansas City, Kansas 66101
(913) 371-2640

Bricklayers and Allied Craftsmen, International Union of
Merlin Taylor, Sr.
815 15th Street, N.W.
Washington, D.C. 20005
(202) 783-3788

Brick and Clay Workers of America, The United
Roy L. Brown, President
3377 West Broad Street
Columbus, Ohio 43204
(614) 275-0286

Broadcast Employees and Technicians, National Association of
Edward M. Lynch, President
5034 Wisconsin Avenue
Suite 1303
Washington, D.C. 20014
(301) 657-8420

Carpenters and Joiners of America, United Brotherhood of
Paul Connelly
Carpenter's Building
101 Constitution Avenue, N.W.
Washington, D.C. 20001
(202) 546-6206

Cement, Lime and Gypsum Workers International Union, United
Thomas Balanoff
7830 West Lawrence Avenue
Chicago, Illinois 60656
(312) 457-1177

Chemical Workers Union, International
Stanley Eller
1655 West Market Street
Akron, Ohio 44313
(216) 867-2444

Clothing and Textile Workers Union, Amalgamated
Eric Frumin
770 Broadway
New York, New York 10003
(212) 255-7800

Communications Workers of America
John Kulstad
1925 K Street, N.W.
Washington, D.C. 20006
(202) 785-6700

Coopers International Union of North America
Ernest D. Higdon
President and Secretary-Treasurer
183 Mall Office Center
400 Sherburn Lane
Louisville, Kentucky 40207
(502) 897-3274

Distillery, Wine and Allied Workers International Union, AFL-CIO/CLC
George J. Oneto, President
66 Grand Avenue
Englewood, New Jersey 07631
(201) 569-9212

Education Association, National
Willard McGuire, President
1201 16th Street, N.W.
Washington, D.C. 20036
(202) 833-4000

Electrical, Radio and Machine Workers, International Union of
William Gary
1126 16th Street, N.W.
Washington, D.C. 20036
(202) 296-1200

Electrical, Radio, and Machine Workers of America, United
Howard Foreman
11 East 51st Street
New York, New York 10022
(212) 753-1960

Electrical Workers, International Brotherhood of
Charles Tupper
1125 15th Street, N.W.
Washington, D.C. 20005
(202) 833-7000

Elevator Constructors, International Union of
Everett A. Treadway, President
Clark Building
Suite 332
5565 Sterrett Place
Columbia, Maryland 21044
(301) 997-9000

Engineers, International Union of Operating
J. C. Turner, President
1125 17th Street, N.W.
Washington, D.C. 20036
(202) 347-8560

Farm Workers of America, AFL-CIO, United
Cesar E. Chavez, President
P.O. Box 62
Keene, California 93531
(805) 822-5571

Fire Fighters, International Association of
Michael Smith
United Unions Building
1750 New York Avenue, N.W.
Washington, D.C. 20006
(202) 872-8484

Firemen and Oilers, International Brotherhood of
John J. McNamara, President
V.F.W. Building, 5th Floor
200 Maryland Avenue, N.E.
Washington, D.C. 20002
(202) 547-7540

Flight Engineers' International Association
William A. Gill, Jr., President
905 16th Street, N.W.
Washington, D.C. 20006
(202) 347-4511

Food and Commercial Workers International Union, United
William H. Wynn, President
1775 K Street, N.W.
Washington, D.C. 20006
(202) 223-3111

Furniture Workers of America, United
Harris Raynor
1910 Airlane Drive
Nashville, Tennessee 37210
(615) 889-8860

Garment Workers of America, United
William O'Donnell, President
200 Park Avenue, South
Suite 1610-1614
New York, New York 10003
(212) 677-0573

Garment Workers Union, International Ladies'
Joseph Danahy
1710 Broadway
New York, New York 10019
(212) 265-7000

Glass and Ceramic Workers
of North America, United
Joseph Roman, President
556 East Town Street
Columbus, Ohio 43215
(614) 221-4465

Glass Bottle Blowers' Association of the United States
and Canada
James E. Hatfield, President
608 East Baltimore Pike
P.O. Box 607
Media, Pensylvania 19063
(215) 565-5051

Glass Workers Union,
American Flint
George M. Parker, President
1440 S. Byrne Road
Toledo, Ohio 43614
(419) 385-6687

Government Employees,
American Federation of
John Albertson
1325 Massachusetts Avenue, N.W.
Washington, D.C. 20005
(202) 737-8700

Grain Millers, American
Federation of
Joseph T. Smisek
4949 Olson Memorial Highway
Minneapolis, Minnesota 55422
(612) 545-0211

Granite Cutters, International
Association of America, The
Joseph P. Ricciarelli, President
18 Federal Avenue
Quincy, Massachusetts 02169
(617) 472-0209

Graphic Arts International
Union
Leonard Adam
1900 L Street, N.W.
Washington, D.C. 20036
(202) 872-7900

Hatters, Cap and Millinery
Workers International Union,
United
John Ross
105 Madison Avenue
New York, New York 10016
(212) 683-5200

Hotel and Restaurant Employees'
and Bartenders' International
Union
Edward T. Hanley, General
President
120 East Fourth Street
Cincinnati, Ohio 45202
(513) 621-0300

Industrial Workers of America,
International Union, Allied
Raymond MacDonald
3520 West Oklahoma Avenue
Milwaukee, Wisconsin 53215
(414) 645-9500

Insurance Workers International
Union, AFL-CIO
Joseph Pollack, President
1017 12th Street, N.W.
Washington, D.C. 20005
(202) 783-1127

Iron Workers, International
Association of Bridge and
Structural
Robert Cooney
400 First Street, N.W.
Washington, D.C. 20001
(202) 783-5848

Jewelry Workers Union, International
Leon Sverdiove, President
8 West 40th Street, No. 501
New York, New York 10018
(212) 244-8793

Laborers' International Union
  of North America
Joseph Short
905 16th Street, N.W.
Washington, D.C. 20006
(202) 737-8320

Laundry and Dry Cleaning
  International Union, AFL-CIO
Russell R. Crowell, President
610 16th Street, No. 421
Oakland, California 94612
(415) 893-1796

Leather Goods, Plastics and
  Novelty Workers Union,
  International
Frank Casale, President
265 West 14th Street, 14th Floor
New York, New York 10011
(212) 675-9240

Leather Workers International
  Union of America
Arthur Cecelski, President
11 Peabody Square
Peabody, Massachusetts 01960
(617) 531-5605

Letter Carriers, National
  Association of
Gustave J. Johnson
100 Indiana Avenue, N.W.
Washington, D.C. 20001
(202) 393-4695

Longshoremen's Association,
  AFL-CIO International
Joseph Leonard
17 Battery Place
Room 1530
New York, New York 10004
(212) 425-1200

Machinists and Aerospace
  Workers, International
  Association of
George Robinson
Machinists Building
1300 Connecticut Avenue, N.W.
Washington, D.C. 20036
(202) 857-5200

Maintenance of Way Employees,
  Brotherhood of
Ole M. Berge, President
12050 Woodward Avenue
Detroit, Michigan 48203
(313) 868-0489

Marine and Shipbuilding Workers
  of America, Industrial Union of
Frank Derwin, President
1126 16th Street, N.W.
Washington, D.C. 20036
(202) 223-0902

Marine Engineers' Beneficial
  Association, National
Jesse M. Calhoon, President
444 North Capitol Street, N.W.
Suite 800
Washington, D.C. 20001
(202) 347-8585

Maritime Union of America,
  National
Shannon J. Wall, President
346 West 17th Street
New York, New York 10011
(212) 924-3900

Mechanics Educational Society
  of America
Alfred J. Smith, National
  President
1421 First National Building
Detroit, Michigan 48226
(313) 965-6990

Metal Polishers, Buffers, Platers
  and Allied Workers
James Siebert, President and
  Secretary-Treasurer
5578 Montgomery Road
Cincinnati, Ohio 45212
(513) 531-2500

Molders and Allied Workers Union, AFL-CIO International
James Wolfe
1225 East McMillan Street
Cincinnati, Ohio 45206
(513) 221-1525

Musicians, American Federation of
Victor W. Fuentealba, President
1500 Broadway
New York, New York 10036
(212) 869-1330

Newspaper Guild, The
Charles A. Perlik, Jr., President
2125 15th Street, N.W.
Washington, D.C. 20005
(202) 296-2990

Novelty Production Workers, International Union of Allied
Julius Isaacson, President
147-49 East 26th Street
New York, New York 10010
(212) 889-1212

Office and Professional Employees International Union
John Kelly, President
265 West 14th Street, No. 610
New York, New York 10011
(212) 675-3210

Oil Chemical and Atomic Workers International Union
Steve Wodka
1126 16th Street, N.W.
Washington, D.C. 20036
(202) 223-5770

Painters and Allied Trades of the United States and Canada, International Brotherhood of
Frank Burkhardt
United Unions Building
1750 New York Avenue, N.W.
Washington, D.C. 20006
(202) 637-0700

Paperworkers International Union, United
Vern McDougal
163-03 Horace Harding Expressway
Flushing, New York 11365
(212) 762-6000

Pattern Makers League of North America
Charles Romelfanger, President
1925 North Lynn Street
Arlington, Virginia 22209
(703) 525-9234

Plasterers' and Cement Masons' International Association of the United States and Canada, Operative
Joseph T. Power, President
1125 17th Street, N.W.
Washington, D.C. 20036
(202) 393-6569

Plumbing and Pipe Fitting Industry of the United States and Canada, United Association of Journeymen and Apprentices of the
Joseph Adam
901 Massachusetts Avenue, N.W.
Washington, D.C. 20001
(202) 628-5823

Police Associations, International Union of
Edward J. Kiernan, President
422 First Street, S.E.
Washington, D.C. 20003
(202) 546-0010

Postal Workers Union,
 AFL-CIO, American
Emmet Andrews, President
817 14th Street, N.W.
Washington, D.C. 20005
(202) 638-2304

Pottery and Allied Workers,
 International Brotherhood of
Lester H. Null, President
P.O. Box 988
East Liverpool, Ohio 43920
(216) 386-5653

Printers, Die Stampers and
 Engravers Union of North
 America, International Plate
Angelo Lo Vecchio, President
1200 S. Court House Road,
 No. 839
Arlington, Virginia 22204
(703) 920-0339

Printing and Graphic Communications Union International
Sol Fishko, President
1730 Rhode Island Avenue,
 N.W.
Washington, D.C. 20036
(202) 293-2185

Professional and Technical
 Engineers, International
 Federation of
Rodney A. Bower, President
1126 16th Street, N.W., No. 111
Washington, D.C. 20036
(202) 223-1811

Pulp and Paper Workers, Association of Western
Farris H. Bryson, President
1430 Southwest Clay
P.O. Box 1735
Portland, Oregon 97207
(503) 228-7486

Radio Association, American
William R. Steinberg, President
270 Madison Avenue
Room 207
New York, New York 10016
(212) 689-5754

Railway, Airline and Steamship
 Clerks, Freight Handlers
 Express and Station Employees,
 Brotherhood of
Fred J. Kroll, President
3 Research Place
Rockville, Maryland 20850
(301) 948-4910

Railway Carmen of the United
 States and Canada, Brother-
 hood
Orville W. Jacobson, President
4929 Main Street
Carmen's Building
Kansas City, Missouri 64112
(816) 561-1112

Railway Employees Department
James E. Yost, President
Room 1212
220 S. State Street
Chicago, Illinois 60604
(312) 427-9546

Railway Supervisors Association,
 American
Frank Ferlin, Jr., President
4250 West Montrose Avenue
Chicago, Illinois 60641
(312) 282-9424

Retail, Wholesale and Department Store Union
Alvin E. Heaps, President
101 West 31st Street
New York, New York 10001
(212) 947-9303

Roofers, Waterproofers and Allied Workers, United Union of
Joseph Bissell
1125 17th Street, N.W.
Washington, D.C. 20036
(202) 638-3228

Rubber, Cork, Linoleum and Plastic Workers of America, United
Lewis Beliczky
URWA Building
87 South High Street
Akron, Ohio 44308
(216) 376-6181

School Administrators, American Federation of
Albert L. Morrison, President
110 East 42nd Street
New York, New York 10017
(212) 697-5111

Seafarers International Union of North America
Paul Hall, President
675 Fourth Avenue
Brooklyn, New York 11232
(212) 499-6600

Service Employee International Union, AFL-CIO
George Hardy, President
2020 K Street, N.W.
Washington, D.C. 20006
(202) 452-8750

Sheet Metal Workers International Association
Edward J. Carlough, President
United Unions Building
1750 New York Avenue, N.W.
Washington, D.C. 20006
(202) 296-5880

Siderographers, International Association of
James C. Small, President
1134 Boulevard
New Milford, New Jersey 07646
(201) 836-9158

Signalmen of America, Brotherhood, Railroad
R. T. Bates, President
601 West Golf Road
Mount Prospect, Illinois 60056
(312) 439-3732

Stage Employees and Moving Picture Machine Operators of the United States and Canada, International Alliance of Theatrical
Walter F. Diehl, President
515 Broadway, No. 601
New York, New York 10036
(212) 730-1770

State, County and Municipal Employees, American Federation of
Donald Wasserman
1625 L Street, N.W.
Washington, D.C. 20036
(202) 452-4800

Steelworkers of America, United
Adolph Schwartz
Five Gateway Center
Pittsburgh, Pennsylvania 15222
(412) 562-2400

Stove, Furnace and Allied Appliance Workers of North America
George E. Pierson, President
2929 South Jefferson Avenue
St. Louis, Missouri 63118
(314) 664-3736

Teachers, American Federation of
Albert Shanker, President
11 Dupont Circle, N.W.
Washington, D.C. 20036
(202) 797-4440

Telegraph Workers, United
D. J. Beckstead, President
701 Gude Drive
Rockville, Maryland 20850
(301) 762-4444

Teamsters, Chauffeurs, Warehousemen and Helpers of America, International Brotherhood of
R. V. Durham
25 Louisiana Avenue, N.W.
Washington, D.C. 20001
(202) 624-6800

Textile Workers of America, United
Francis Schaufenbil, President
420 Common Street
Lawrence, Massachusetts 08140
(617) 686-2901

Tile, Marble, Terrazzo Finishers and Shopmen International Union, AFL-CIO
Pascal Di James, President
801 North Pitt Street, No. 116
Alexandria, Virginia 22314
(703) 549-3050

Train Dispatchers Association, American
B. C. Hilbert, President
1401 South Harlem Avenue
Berwyn, Illinois 60402
(312) 795-5656

Transit Union, Amalgamated
Dan V. Maroney, Jr., President
5151 Wisconsin Avenue, N.W.
3rd Floor
Washington, D.C. 20016
(202) 537-1645

Transport Workers Union of America
William G. Lindner, President
1980 Broadway
New York, New York 10023
(212) 873-6000

Transportation Union, United
Al H. Chesser, President
14600 Detroit Avenue
Cleveland, Ohio 44107
(216) 228-9400

Typographical Union, International
Joseph Bingel, President
P.O. Box 157
Colorado Springs, Colorado 80901
(303) 636-2341

Upholsterers' International Union of North America
Sal B. Hoffmann, President
25 North Fourth Street
Philadelphia, Pennsylvania 19106
(215) 923-5700

Utility Workers Union of America
Marshall M. Hicks
815 16th Street, N.W.
Suite 605
Washington, D.C. 20006
(202) 347-8105

Woodworkers of America, International
R. Denny Scott
1622 North Lombard Street
Portland, Oregon 97217
(503) 285-5281

Yardmasters of America, Railroad
A. T. Otto, Jr., General President
1411 Peterson Avenue
Room 201-202
Park Ridge, Illinois 60068
(312) 696-2510

SELECTED AFL-CIO DEPARTMENTS

Building and Construction
 Trades Department
James Lapping
(202) 347-1461

Building and Construction
 Trades Safety and Health
 Committee
Robert Cooney, Chairman
400 First Street, N.W.
Washington, D.C. 20001
(202) 783-5848

Education, Department of
Edward Czarnecki
(202) 637-5141

Industrial Union Department
Sheldon Samuels
(202) 393-5581

Legislation, Department of
Ray Denison, Director
(202) 637-5075

Maritime Trades Department
Paul Hall
(202) 628-6300

Metal Trades Department
Paul J. Burnsky, President
(202) 347-7255

Occupational Safety and Health,
 Department of
George Taylor, Director
(202) 637-5175

Public Employee Department
John McCart
(202) 393-2820

FEDERAL ADVISORY COUNCIL ON OCCUPATIONAL SAFETY AND HEALTH, JANUARY 1980

Chairperson

Assistant Secretary of Labor
 for Occupational Safety and
 Health
Washington, D.C. 20210
(202) 523-9362

Members Term to 12/31/80

Jerry A. Jones
General Manager
Accident Prevention Division
U.S. Postal Service
Room 9912
475 L'Enfant Plaza, West
Washington, D.C. 20260
(202) 245-4686

John A. McCart
Executive Director

Public-Employee Department,
 AFL-CIO
815 16th Street, N.W.
Washington, D.C. 20006
(202) 393-2820

George Marienthal
Deputy Assistant Secretary of
 Defense (Energy, Environ-
 ment and Safety)
Room 3B252, The Pentagon
Washington, D.C. 20301
(202) 695-0221

Vincent C. Sombrotto, President
National Association of Letter
 Carriers, AFL-CIO
100 Indiana Avenue, N.W.
Washington, D.C. 20001
(202) 393-4695

Members Term to 12/31/81

Robert L. Crum
International Brotherhood of
 Electrical Workers, AFL-CIO
1125 15th Street, N.W.
Washington, D.C. 20005
(202) 833-7101

Emmet Andrews
General President
American Postal Workers Union
817 14th Street, N.W.
Washington, D.C. 20005
(202) 638-2308

John Meese
National Coordinator
Government Employees Department
International Association of
 Machinists
1300 Connecticut Avenue, N.W.
Washington, D.C. 20036
(202) 857-5200

Dr. Joan S. Wallace
Assistant Secretary for
 Administration
U.S. Department of Agriculture
14th & Independence Avenue,
 S.W.
Washington, D.C. 20250
(202) 447-3291

Vacancy (Mgt.)

Members Term to 12/31/82

Kenneth T. Blaylock
President
American Federation of Government Employees, AFL-CIO
1325 Massachusetts Avenue,
 N.W.
Washington, D.C. 20005
(202) 737-8700

Anthony F. Ingrassis, Vice-Chairperson
Assistant Director
Labor-Management Relations
Office of Personnel Management
Washington, D.C. 20415
(202) 632-4468

George H. R. Taylor
Executive Secretary
AFL-CIO Standing Committee
 on Occupational Safety and
 Health
815 16th Street, N.W.
Washington, D.C. 20006
(202) 637-5175

(2) Vacancy (Mgt.)

FACOSH Consultant

Marshall LaNier, Director
Division of Technical Services
National Institute of Occupational Safety and Health
4676 Columbia Parkway
Cincinnati, Ohio 45226
(513) 684-8302

FACOSH Executive Director

Annie Asensio
Office of Federal Agency Safety
 and Health Programs
Department of Labor, OSHA
200 Constitution Avenue, N.W.
Room N3423
Washington, D.C. 20210
(202) 523-8677

APPENDIX I: A GUIDE TO WORKER EDUCATION MATERIALS IN OCCUPATIONAL SAFETY AND HEALTH

PREFACE

This guide lists occupational safety and health materials produced by labor unions. Any type of material—written, audio-visual, textual, or in outline format—suitable for worker training and education was sought. This guide has tried to list as many useful resources as possible, but it can in no way claim to be comprehensive.

All prices quoted are current as of April 1980. If no price is indicated, the item is either free of charge or carries a charge unknown to us. Please check with the source about the cost.

The source index contains most of the descriptive information regarding format, length, publication date, price, and contents. More detailed written materials are indicated with underlining; audiovisual materials are set off with quotation marks. Less detailed written materials (for example, factsheets) have no distinguishing markings.

The subject index cross references items found in the author index and places them in one or more of 26 subject headings. Abbreviations used in this section can be found in the List of Abbreviations. Other indexes list newsletters, audiovisual, and foreign language materials.

The guide was prepared by Brett Hill, Project Director, with the support of other staff members of the Urban Environment Conference. Thanks are due to the staff members of the unions listed here for their cooperation in completing this guide. For further information about the materials in this survey contact:

Urban Environment Conference
666 11th Street, N.W., Suite 1001
Washington, D.C. 20001
(202) 638-3385

SOURCE INDEX CONTENTS

American Federation of Labor-Congress of Industrial Organizations
    Building and Construction Trades Department
    Food and Beverage Trades Department
    Industrial Union Department
    Occupational Safety and Health; Department of
    Public Employee Department
Asbestos Workers; International Association of Heat and Frost Insulators and
Automobile, Aerospace and Agricultural Implement Workers of America; International Union, United
Automobile, Aerospace and Agricultural Implement Workers of America; International Union, United—District 65 (formerly District 65, Distributive Workers of America)

Bakery, Confectionery and Tobacco Workers International Union
Boilermakers, Iron Ship Builders, Blacksmiths, Forgers and Helpers; International Brotherhood of
Boilermakers, Iron Ship Builders, Blacksmiths, Forgers and Helpers; International Brotherhood of—Local 169
Boilermakers, Iron Ship Builders, Blacksmiths, Forgers and Helpers; International Brotherhood of—Local 802

Chemical Workers Union; International
Clothing and Textile Workers Union; Amalgamated
Communications Workers of America

Electrical, Radio and Machine Workers of America; International Union of
Electrical, Radio and Machine Workers of America; International Union of—Local 201
Electrical, Radio and Machine Workers of America; United

Fire Fighters; International Association of
Furniture Workers of America; United

Graphic Arts International Union

Industrial Workers of America; International Union Allied

Longshoremen's and Warehousemen's Union; International

Machinists and Aerospace Workers; International Association of
Meat Cutters; Amalgamated/Retail Food Store Employees—Local 342
Molders and Allied Workers Union; International

APPENDIX I / 237

Office and Professional Employees International Union
Oil, Chemical and Atomic Workers International Union

Painters and Allied Trades of the United States and Canada;
   International Brotherhood of
Paperworkers International Union; United
Plasterers' and Cement Masons' International Association of the
   United States and Canada; Operative
Plumbing and Pipe Fitting Industry of the United States and
   Canada; United Association of Journeymen and Apprentices
   of the

Railway, Airline and Steamship Clerks, Freight Handlers,
   Express and Station Employees; Brotherhood of
Roofers, Damp and Waterproof Workers Association
Rubber, Cork, Linoleum and Plastic Workers of America; United

Service Employees International Union
State, County and Municipal Employees; American Federation of
State, County and Municipal Employees; American Federation
   of—District Council 37
Steelworkers of America; United

Woodworkers of America; International

SUBJECT INDEX CONTENTS

General
    General Works
    Hazard Identification and Abatement
    Medical Tests and Information
    OSHA—Guides to Standards and Procedures
    Personal Protective Equipment
    Resource Guides
    Sample Forms and Letters
    Union Activities and Contract Language
    Women's Occupational Safety and Health
    Workers' Compensation

Hazards
    Accidents and Accident Causation
    Asbestos
    Chemical
    Dusts
    Fires and Explosions

238 / SAFETY AND HEALTH CATALOGUE

    Lead
    Noise
    Physical and Safety

Occupations
    Construction
    Foundry
    Grain
    Hospital and Health Care
    Office
    Public Sector
    Shipyard
    Welding

OTHER INDEXES CONTENTS

Newsletters

Audiovisual Materials

Foreign Language Materials

LIST OF ABBREVIATIONS

ACTWU - Amalgamated Clothing and Textile Workers Union
AFL-CIO - American Federation of Labor-Congress of Industrial Organizations
AFSCME - American Federation of State, County and Mulitipal Employees
AIW - International Union Allied Industrial Workers of America
Asbestos Workers - International Association of Heat and Frost Insulators and Asbestos Workers
BC&T - Bakery, Confectionery and Tobacco Workers International Union
Boilermakers - International Brotherhood of Boilermakers, Iron Ship Builders, Blacksmiths, Forgers and Helpers
BRAC - Brotherhood of Railway, Airline and Steamship Clerks, Freight Handlers, Express and Station Employees
CWA - Communications Workers of America
DC 37 - American Federation of State, County and Municipal Employees District Council 37
District 65 - District 65, International Union, UAW
GAIU - Graphic Arts International Union

IAFF - International Association of Fire Fighters
IAM - International Association of Machinists and Aerospace Workers
IBPAT - International Brotherhood of Painters and Allied Trades of the United States and Canada
ICWU - International Chemical Workers Union
ILWU - International Longshoremen's and Warehousemen's Union
IUD - AFL-CIO Industrial Union Department
IUE - International Union of Electrical, Radio and Machine Workers
IWA - International Woodworkers of America
Molders - International Molders and Allied Workers Union
OCAW - Oil, Chemical and Atomic Workers International Union
OPEIU - Office and Professional Employees International Union
PED - AFL-CIO Public Employee Department
Plasterers - Operative Plasterers' and Cement Masons' International Association
RDWW - Roofers, Damp and Waterproof Workers Association
SEIU - Service Employees' International Union
UA; United Association - United Association of Journeymen and Apprentices of the Plumbing and Pipefitting Industry of the United States and Canada
UAW - International Union, United Automobile, Aerospace and Agricultural Implement Workers of America
UE - United Electrical, Radio and Machine Workers of America
UFWA - United Furniture Workers of America
UPIU - United Paperworkers International Union
URW - United Rubber, Cork, Linoleum and Plastic Workers of America
USWA - United Steelworkers of America

SOURCE INDEX

AFL-CIO Building and Construction Trades Department
Safety and Health Education and Training Program
815 16th Street, N.W.
Washington, D.C. 20006
(202) 347-1461

Book
  <u>Guidebook to Occupational Safety and Health</u> (417 pp., 1977)
    Comprehensive guide to the Occupational Safety and Health Act and OSHA administrative procedures. Published by Commerce Clearing House.

Courses (3-4 hours)
The Building and Construction Trades Department offers the following course modules:

- Introduction to Construction Hazard Control
- Overview of Hazards in the Construction Industry
- Health Hazards and Personal Protective Equipment
- Excavation and Trenching Safety: Soil Characteristics
- Electrical Hazard Control
- Site Inspection Techniques
- Techniques of Accident Investigation
- Materials Handling and Crane Safety
- Construction Hazard Analysis
- Communication Techniques
- Training and Educational Methods

AFL-CIO, Food and Beverage Trades Department (FBTD)
Occupational Safety and Health Training Program
815 16th Street, N.W.
Washington, D.C. 20006
(202) 347-2640

Booklets
Combating Hazards on the Job: A Workers' Guide (42 pp., 1979)
This booklet outlines hazards faced by FBTD affiliate unions and possible solutions including OSHA remedies, health and safety committees, and contract clauses.

The Exploding of the Grain Belt: An Action Program for the American Federation of Grain Millers (64 pp., 1979)
This booklet describes major safety and health problems in the grain industry and outlines corrective measures including filing OSHA complaints, forming union health and safety committees, and contract clauses.

Grain Elevators Bulk Handling Facilities—NFPA 61B (29 pp., 1973)
Consensus standards for the prevention of fire and dust explosions in grain elevators and Bulk Grain Handling Facilities. Published by the National Fire Protection Association.

Newsletter Issue
F & B Topics (quarterly, 8 pp.) Note: Subscriptions to FBTD members only.
"FBTD Special Report: Safety and Health on the Job" 3, no. 2 (Summer 1979)

Grain Elevator Health and Safety Factsheets
- Fires and Explosions
    - Bearing Hazards
    - Bucket Elevator Hazards and Design
    - Conveyor Hazards and Design
    - Distribution Floor and Bin Design
    - Dust Collection Design and Operation
    - Electrical Equipment Standards
    - Explosion Suppression and Inerting
    - Explosion Venting
    - General Fire Prevention and Emergency Procedures
    - Static Electricity Removal
- Fumigant Hazards
    - Application Methods for Liquid Fumigants
    - Application Methods for Phostoxin
    - Carbon Disulfide
    - Carbon Tetrachloride
    - Ethylene Dibromide
    - Ethylene Dichloride
    - Less Frequently Used Substances (Chloropicrin, Calcium Cyanide, Ethyl and Methyl Formate, and Sulfur Dioxide)
    - Malathion
    - Methyl Bromide
    - Phostoxin
    - Rodenticides
    - Safety and Health Rules for Fumigant Use
- Grain Industry Diseases
    - Asthma
    - Bronchitis
    - Farmers' Lung
    - Grain Fever
    - Grain Itch
    - Malt Workers' Lung
- General Safety
    - Fall Protection
    - Lockout and Tagout Procedures
    - Machine Guarding
    - Manlift Hazards and Design
    - Safe Bin Entry
- General Health
    - Control Measures for Cleaning up the Workplace
    - Hexane Health and Explosion Hazards
    - Noise
    - Respirators

OSHA Rights and Responsibilities
Electing Party Status
Calling for an OSHA Inspection
Employer Responsibilities under the OSHA LAW
NIOSH
Sample Safety & Health Contract Language

Film
"More Than a Paycheck" (28 min.)
Explores the cancer risk faced by industrial workers. Produced by George Washington University under contract with the Occupational Safety and Health Administration.

AFL-CIO, Industrial Union Department
815 16th Street, N.W.
Washington, D.C. 20006
(202) 393-5581

Pamphlet
Work Without Fear (8 pp., 1973)
An early pamphlet explaining provisions of the Occupational Safety and Health Act, standards set under the act, worker rights, and questions with which workers can evaluate their workplaces.

Newsletter
IUD Facts and Analysis (1-2 pp., approx. monthly)
Highlights current events in the occupational safety and health field.

AFL-CIO, Department of Occupational Safety and Health
815 16th Street, N.W.
Washington, D.C. 20006
(202) 637-5175

Article
Women Workers: Hazards on the Job (6 pp., 1978)
Describes hazards of jobs employing large numbers of women, for example, hospital workers, laundry and dry cleaning workers, and clerical workers. Also addresses the issue of protective discrimination by which women are barred from jobs involving exposure to substances posing a potential risk to an unborn child.

AFL-CIO, Public Employee Department
815 16th Street, N.W.
Washington, D.C. 20006
(202) 393-2820, 2821

Booklet
Occupational Safety and Health: A Promise Unfulfilled for Public Employees (42 pp., 1977)
 Evaluates standards and practices under the Occupational Safety and Health Act that affect public-sector employees. Includes a statistical review of public sector injury rates and contractual provisions in public-sector contracts.

International Association of Heat and Frost Insulators
and Asbestos Workers (AFL-CIO)
1300 Connecticut Avenue, N.W., Suite 511
Washington, D.C. 20036
(202) 785-2388

Pamphlet
The Law: What It Demands in the Process of Stripping Asbestos (2 pp., 1979)
 Describes safe work practices for the removal of asbestos.

International Union, United Automobile, Aerospace and
Agricultural Implement Workers of America
Social Security Department
8000 East Jefferson Avenue
Detroit, Michigan 48214
(313) 926-5000

Booklets
Control of Lead in Battery Making (12 pp., 1979)
Foundry Fatality Prevention (32 pp., 1977)
 Presents a series of descriptions of fatal accidents in the foundry industry and suggests preventive procedures for each situation. Includes French translation.
The Hazards of Lead and How to Control Them (17 pp., 1979)
 Describes health effects and control programs.
How to Evaluate On-The-Job Health Hazards (12 pp.)
 Describes hazard recognition procedures, measuring devices and procedures, and sets up a methodology by which to analyze health hazards. Original publisher: National Safety Council.

Lead Hazards in Vehicle Assembly and Repair Operations: Body Solder (10 pp., 1977)
  Outlines procedures for lead control in body solder operations.
Lung Function Tests (8 pp., 1977)
  Describes common lung function tests and how they can be used to detect lung disease.
Machinery Lockout Procedures (13 pp., 1979, $1.25, order number SSD5)
  Details lockout procedures, sample methods for specific types of controls, and suggests a minimally satisfactory lockout checklist.
Noise Control: A Worker's Manual (49 pp., 1979, $1.25, order number SSD4)
  Describes basic principles of noise control including design principles and noise control for specific machines, and describes the effects of excessive noise on hearing and the human body.
Occupational Safety and Health Hazard Guide (24 pp., $.50)
  Reference guide to common job hazards.
Skilled Trades Fatality Prevention (19 pp., 1976)
  Presents 13 case studies involving fatal accidents to UAW members. Each accident is analyzed for causes and remedies are suggested.
What Every UAW Representative Should Know About Health and Safety (32 pp., 1979, $1.00, order number SSD449)
  A general overview of common job hazards, protective laws and agencies, hazard recognition, and health and safety resources.

Factsheets (2-4 pp.)
  Common Toxic Substances and Their Effects (order number SSD2)
  Controlling Deadly Vapors
  How to Crack the Company's Code: Getting Names of Workplace Chemicals
  How to Get Information from OSHA
  How to Prevent Skin Disease Caused by Cutting Fluids
  How to Use the Lead Standard: A Local Union Action Program
  Lead: A Worker's Guide to Checking Exposure to Lead
  Polyurethane: Job Health Hazard?
  Strapped to a Mask? Here are the Rules for Respirators
  A Union Representative's Guide to Occupational Safety and Health (order number SSD1)

APPENDIX I / 245

Newsletter
  Occupational Health and Safety Newsletter (4 pp., bimonthly, $3.50/year)
    Highlights recent legislative and regulatory developments and frequently contains feature articles on the recognition and control of specific hazards.

Training Curricula
  Developed by the UAW Education Department for the UAW Independents, Parts and Suppliers (IPS) Department. Instructional units include OSHA, Hazardous Substances, Ventilation, and Noise.

Foreign Language Publications (translations by and available from UAW International Affairs Department, 1757 N Street, N.W., Washington, D.C. 20036, (202) 828-8500).

Spanish

Booklet
  Lo Que Todo UAW Representante Debe Saber Sobre la Salud y la Seguridad (63 pp., 1979)
    Translation of What Every UAW Representative Should Know About Health and Safety.

Factsheets
  Arsénico (2 pp.)
    Presents arsenic hazards, exposure limits, and control methods.
  Asbestos (1 pp.)
  Como Prevenir Enfermedades de la Piel Causadas por los Fluidos de Cortar (2 pp.)
    Presents cutting fluid hazards and personal and engineering controls.
  Dermatitis (2 pp.)
  Diseño para una Ventilación de Extracción Local (2 pp.)
    Designs for a local exhaust ventilation system.
  El Effecto de Calor (2 pp.)
    Describes heat hazards, exposure limits, and control methods.
  ¿Está Su Cara Liada a una Mascara? Estas Son las Reglas que Rigen a los Respiradores (3 pp., 1976)
    Describes a respirator use program.
  Fibras de Vidrio (2 pp.)
    Discusses fiberglass hazards, control, and personal protection.

Fluorocarburos (2 pp.)
: Describes fluorocarbons and their control.

El Formaldehido (1 p.)
: Describes the health risks and control of formaldehyde.

Monóxido de Carbono (2 pp.)
: Describes hazards of carbon monoxide.

Los Peligros que Representa el Plomo y como Controlarlos (17 pp., 1977)
: Describes bodily effects and control of lead hazards.

Silicosis (3 pp.)

Tricloroetileneo (3 pp.)
: Describes trichloroethylene hazards and control.

Ventilación (8 pp.)
: Discusses general and local exhaust ventilation systems.

Portuguese

Factsheets

Cloro (2 pp.)
: Discusses hazards of chlorine gas and control methods.

Dermatite (2 pp.)
: Outlines causes, prevention, and treatment of dermatitis.

French

Booklet

La Prevention des Accidents Mortels dans les Fonderies (16 pp., 1977)
: French translation of Foundry Fatality Prevention

International Union, United Automobile, Aerospace and
Agricultural Implement Workers of America, District 65
(Formerly District 65, Distributive Workers of America)
Occupational Safety and Health Department
13 Astor Place
New York, New York 10003
(212) 673-5120

Factsheets
Checklist of Possible Health and Safety Problems in Your Shop (2 pp.)
Information on Occupational Health and Safety (2 pp.)

Contract Language
Health and Safety Clause (2 pp.)

APPENDIX I / 247

Bakery, Confectionery and Tobacco Workers
International Union (AFL-CIO)
1828 L Street, N.W., Suite 900
Washington, D.C. 20036
(202) 466-2500

Newsletter
BC&T Report (4-8 pp., monthly)
  Usually contains one or more OSHA-related articles on new publications, specific hazards, legal briefs, or local union activities.

Newspaper
BC&T News (8 pp., monthly)
  Covers current events in the occupational safety and health field.

International Brotherhood of Boilermakers, Iron Ship Builders, Blacksmiths, Forgers and Helpers (AFL-CIO)
570 New Brotherhood Building
Kansas City, Kansas 66101
(913) 371-2640

Booklets
How to Put the Occupational Safety and Health Act to Work: A Guideline for Stewards (32 pp.)
  Outlines basic provisions of OSHA, employee and employer rights, filing a complaint, enforcement procedures, and the operation of the Review Commission. Also includes a description of the Toxic Substances Control Act.
The Safety Committee and Collective Bargaining: New Concepts (29 pp., 1979)
  Discusses the need for strong safety committees, how to organize one, sample contract language, and in-plant inspection procedures.
The Uncharted Seas of Shipyard Safety and Health (12 pp., 1977)
  Describes common shipyard safety and health hazards, the role of OSHA, and methods of hazard abatement.

International Brotherhood of Boilermakers, Iron Ship Builders, Blacksmiths, Forgers and Helpers (AFL-CIO), Local 169
Safety Department

5936 Chase Road
Dearborn, Michigan 48126
(313) 584-8520

Booklets
Precautions and Safe Practices for Boilermaker Welders (12 pp., 1980)
   Covers protective equipment, preparation for welding (fire and explosion prevention), and health hazards of welding. Also available from Scott Tobey, Labor Program Service, Michigan State University, South Kedzie Hall, East Lansing, Michigan 48824.
Radiation and Radiation Protection (8 pp., 1979)
   This easy-to-read guide explains basic concepts in radiation: types of radiation, dosage, exposure limits, and protection from exposure.
Guide to Safe Rigging Practices
   Wallet-sized guide to safe rigging practices, timber and cable strength, and other rigging data.

International Brotherhood of Boilermakers, Iron Ship Builders, Blacksmiths, Forgers and Helpers (AFL-CIO), Local 802
301 East Third Street
Chester, Pennsylvania 19016
(215) 872-3635

Pamphlets
Health Problems for Lifting and Carrying (7 pp., 1979)
   Describes common injuries, causes, and prevention of back and other lifting and carrying injuries, and includes exercises for lower back pain.
Safe Staging on Ships (2 pp., 1979)
   Describes OSHA requirements for staging in the shipbuilding industry.

International Chemical Workers Union (AFL-CIO)
Occupational Health and Safety Department
1655 West Market Street
Akron, Ohio 44313
(216) 867-2444

Manual
Health and Safety Manual

As currently planned, this manual is to cover health and safety committee formation and duties, plant survey and record keeping, strategies for hazard abatement, bargaining, and a resource guide.

Hazard Identification and Abatement Guides
  Health Hazard Inventory (4 pp.)
    Used to analyze workplace reporting, monitoring, medical surveillance, and worker education activities.
  Health Hazard Questionnaire (4 pp.)
    Used to identify chemical raw materials and products, intermediates, and physical agent hazards.
  OSHA General Industry Standards (31 pp.)
    Index to the general industry standards (29 CFR 1910).
  OSHA Inspection Check List (4 pp.)
    Used to record information on an OSHA inspection.
  Respiratory Protection Devices (9 pp.)
    Outlines types of respirators, fit, training, and maintenance. Includes American National Standards Institute recommendations for specific uses of different types of eye and face protectors, and a guide to the use of gloves.

Policies and Contract Language
  General Policy of the ICWU on Matters of Health and Safety (3 pp.)
  ICWU Policy on Reproductive Effects of Hazardous Materials (3 pp.)
  Suggested Health and Safety Language (5 pp.)

Factsheets
  Beat the Heat (2 pp., 1979)
  Fluorides (2 pp.)
  Pesticides (2 pp., 1979)
  Polyurethane (1 p., 1977)
  Shift Work and Its Consequences (1 p., 1979)

Sample Letters and Forms (to be included in Health and Safety Manual)
  Closing Conference Demand
  Complaint Against State Program Administration
  Contest of Abatement Date (Review Commission)
  Demand for Bargaining and Request for Company Information—
    Absenteeism
  Demand for Bargaining and Request for Company Information—
    Female Exclusion Policy

250 / SAFETY AND HEALTH CATALOGUE

    Demand for Bargaining and Request for Company Information—Medical Examinations
    11(c) Discrimination Complaint (OSHA)
    Freedom of Information Act Request
    Imminent Danger (OSHA)
    Inspection Request (OSHA)
    Request for Company Information—General Health and Safety
    Request for Party Status (Review Commission)

Newspaper
    The Chemical Worker (12 pp., monthly, $12/year)
        Frequently contains articles on ICWU health and safety activities.

Amalgamated Clothing and Textile Workers Union (AFL-CIO)
Department of Occupational Safety and Health
770 Broadway
New York, New York 10003
(212) 777-3600

Booklets
    ACTWU Local Union Health and Safety Action Manual (30 pp., 1979)
        Covers formation and activities of the safety and health committee, grievance handling, bargaining, OSHA complaints, and worker education. Contains hazard investigation questionnaire, sample OSHA form 200, hazard correction checklist, sample OSHA complaint, sample NIOSH health hazard evaluation form, checklist of local union actions in connection with OSHA complaints and inspections, and a list of educational materials.
    Cotton Dust Control Manual (20 pp., 1980)
        Thoroughly explains the OSHA cotton dust standard. Includes highlights, union activities, use of respirators, exposure monitoring, education and training, medical surveillance, compensation, and the complete text of the standard.

Factsheets
    ACTWU Actions to Protect Worker Safety and Health (3 pp., 1979)
    ACTWU Checklist of Local Union Actions in Connection with OSHA Complaints and Inspections (2 pp.)
    ACTWU Checklist on Hazard Correction (2 pp.)

APPENDIX I / 251

ACTWU-OSHA Training Program
    Goals (1 p.)
    Discussion Outline (12 pp., 1979)
ACTWU Questionnaire for Hazard Investigation (2 pp.)
Asbestos and Pipefitters (1 p., 1980)
Audiometric Monitoring (Hearing Tests) and the Meaning of Test Results (3 pp., 1979)
Cracking Codes on Company Chemicals (1 p., 1979)
Design of Local Exhaust Ventilation (2 pp., 1979)
Earplugs (2 pp., 1979)
Educational Materials (2 pp., 1979)
Fire Brigades (2 pp., 1979)
Glossary of Health Hazard Terms (4 pp., 1979)
Health Hazard as Grounds for Refusal to Operate Grinder (1 p., 1979)
How to Use OSHA Form 200 (2 pp., 1978)
How to Use the OSHA Standards (1 p.)
Local Union Health and Safety Records (1 p., 1979)
Noise Case Study (3 pp.)
OSHA Forms
    Material Safety Data Sheet (2 pp.)
    Supplementary Record of Occupational Injuries and Illnesses (2 pp.)
OSHA Guidelines on Hearing Conservation Program (1 p.)
OSHA Lead Standard—Summary of Medical Surveillance and Job-Transfer Sections (3 pp., 1979)
OSHA Respirator Regulations Guide (5 pp., 1979)
OSHA's Asbestos Standard Guide (4 pp., 1979)
Safety Committee Request Guidelines (1 p., 1979)
Unsafe Acts as a Cause of Accidents (2 pp., 1980)
Workers' Compensation (2 pp.)
Workers' Compensation Facts for Synthetic Division Members (2 pp.)

Newspaper
    <u>ACTWU Labor Unity</u> (approx. 24 pp., monthly, $1.50/year)
    Frequently highlights union activities in occupational health and safety.

Slide/Tape Show
    "Our Lives, Our Rights" (15 min., color, 1978)
    Presents cotton dust and other hazards of the textile industry. Describes potential worker action to abate these hazards.

Communications Workers of America (AFL-CIO)
  Occupational Safety and Health Program
  1925 K Street, N.W.
  Washington, D.C. 20006
  (202) 785-5857

Factsheets
  Arsenic (4 pp.)
  Asbestos (7 pp.)
  Carbon Monoxide (6 pp.)
  Forming CWA Local Occupational Safety and Health Committees
    (4 pp., 1980)
  Freon (2 pp.)
  Lead (5 pp.)
  Polychlorinated Biphenyls (6 pp.)
  Polyurethanes and Isocyanates (7 pp.)
  Vinyl Chloride (6 pp., 1980)
  Visual Display Terminals (5 pp.)

International Union of Electrical, Radio and Machine Workers
  (AFL-CIO)
  1126 16th Street, N.W.
  Washington, D.C. 20036
  (202) 296-1200

Newsletter
  Health and Safety Bulletin (4 pp., 3-6 issues/year)
    The newsletter reports on recent OSHA activities and new
    publications. (Note: Back issues or subscriptions available
    only through special arrangement.)

International Union of Electrical, Radio and Machine Workers,
  Local 201 (AFL-CIO)
  100 Bennett Street
  Lynn, Massachusetts 01904
  (617) 598-2760

In Production

Pamphlets
  Cutting Oils
  Welding

United Electrical, Radio and Machine Workers of America
11 East 51st Street
New York, New York 10022
(212) 753-1960

Booklet
UE Manual on Foundry Health and Safety (11 pp., 1979)
  Contains sections on hazard identification and health and safety control procedures.

Newsletter
UE News (biweekly, $4.00/year)
  Features "Health on the Job," a regular column on hazards and hazard abatement in the electrical manufacturing and machine industries. Also contains occasional feature and news articles on occupational safety and health.

International Association of Fire Fighters (AFL-CIO)
1750 New York Avenue, N.W.
Washington, D.C. 20006
(202) 872-8484

Newsletter
The International Fire Fighter (monthly, 20 pp.)
  Contains monthly articles on safety and health.

Miscellaneous Materials
  The occupational safety and health department has extensive files on personal protective equipment.

United Furniture Workers of America (AFL-CIO)
UFWA National Job Training and Counseling Program
1910 Air Lane Drive
Nashville, Tennessee 37210
(615) 889-8860

Booklet
Watch Out! A Handbook of Safety and Health for Furniture Workers (48 pp., 1979)
  Summarizes work hazards in the furniture industry (including wood, metal, and upholstered furniture, and bedding), formation and use of safety and health committees, worker rights and employer duties under the OSHAct, OSHA complaint and inspection procedures, and available resources.

Graphic Arts International Union (AFL-CIO)
Safety and Health Awareness and Action Program
1900 L Street, N.W.
Washington, D.C. 20036
(202) 872-7928

Training Manual
OSHA/SHAPE Program (330 pp., 1980)
  Each session, designed to last three hours, consists of instructor's outline, student handouts, and accompanying audiovisual materials. Contains sessions on the following topics:
    Introduction to Safety and Health Hazards in Industry and Graphic Arts
    Electrical Hazards, Fires, Walking and Working Surfaces
    Noise
    Toxic Chemicals
    Ergonomics
    Job Stress
    Employee-Employer Rights and Responsibilities under OSHA
    Action Planning

Booklet
Noise Reduction in Printing Plants (11 pp., 1978)
  Discusses noise problems in printing plants, ear protectors, workplace redesign, and insulation. Also contains a general discussion of noise and a glossary of terms. Published by the Swedish Graphics Working Environment Committee.

International Union Allied Industrial Workers of America (AFL-CIO)
Safety and Health Project
AIW Building
3520 West Oklahoma Avenue
Milwaukee, Wisconsin 53215
(414) 645-9500

Training Manual
AIW Safety and Health Training Manual (130 pp., 1979)
  A collection of factsheets on safety and health hazard recognition, control measures, and collective bargaining.

International Longshoremen's and Warehousemen's Union
1188 Franklin Street
San Francisco, California 94109
(415) 775-0533

Booklet
Pacific Coast Marine Safety Code (90 pp., 1979)
  Constitutes the official safety standards for the Pacific Coast Marine Industry.

Contract Language
Pacific Coast Longshore Contract Document 1978-81 (200 pp., 1978)
  Sections 11 and 16 set down employer and union safety and health responsibilities.

International Association of Machinists and Aerospace Workers (AFL-CIO)
Department of Occupational Safety and Health
1300 Connecticut Avenue, N.W.
Washington, D.C. 20036
(202) 857-5262

Book
Help for the Working Wounded (250 pp., rev. 1980, $1.00)
  Presents questions and answers on chemical and other hazards, medical screening, and industrial hygiene.

Booklet
IAM Guide for Safety and Health Committees (24 pp.)
  A guide to organizing safety and health committees. Details committee structure and responsibilities, outlines the OSHAct, and suggests committee recordkeeping and educational programs.

Factsheets—Hazards
  Asbestos (6 pp.)
  Benzene (2 pp.)
  Beryllium (2 pp.)
  Health Hazards Associated with Diesel Engine Emissions (3 pp.)
  Industrial Solvents (3 pp.)
  Lead (5 pp.)—includes update on status of OSHA lead standard
  Noise (5 pp.)
  Nonionizing Radiation (3 pp.)
  Radiation (4 pp.)

Factsheets—Legal Issues
  Barlow Decision (2 pp.)—complete text of Supreme Court decision attached
  Bureau of Motor Carrier Safety Regulations on Employee Safety and Health in the Operation, Maintenance, and Loading and Unloading of Motor Vehicles (2 pp.)
  Status of State Plans (1 p., 1979)
  Walk-Around Violations (2 pp.)
  Whirlpool Decision (1 p.)—complete text of Supreme Court decision attached.

Newsletter
  IAM Safety Gram (2 pp., quarterly)
    Reports new OSHA standards, available resources, and other current events in the occupational safety and health field.

Films
  "More Than a Paycheck" (21 min., color, 16 mm)
    IAM version of the OSHA-sponsored film that surveys a number of workplace settings and stresses the safe use of toxic chemicals and recognizable hazardous substances through proper job and equipment design. The film demonstrates that worker safety and job security are complementary.
  "Paul Jacobs and the Nuclear Gang" (60 min., color, 16mm)
    An investigation of nuclear testing in the 1950s and the health effects of exposure to nuclear radiation.

In Production
  Federal Executive Order 12196
    Discusses the new executive order on federal agency occupational safety and health programs and its impact on IAM members.
  Proceedings of IAM National Safety and Health Conference (1980)

Amalgamated Meat Cutters/Retail Food Store Employees Union,
  Local 342 of Greater New York (AFL-CIO)
(Amalgamated Meat Cutters and Butcher Workmen of North America)
186-16 Hillside Avenue
Jamaica, New York 11432
(212) 479-5000

Film
"I Never Had an Accident in My . . ." (18 min., color, sound, 16 mm, 1975)
Produced in cooperation with OSHA, the film covers five major causes of accidents—lifting, band saws, hand knives, meat wrapping, and slips and falls—provisions of the OSHAct, and a labor-management safety and health meeting. (Available for $108.25/purchase, $25/3-day rental from National Audiovisual Center, Washington, D.C. 20409, (301) 763-1896. Specify Order No. 009871.)

Slide/Tape Show
Local 342 Safety Training Program
Covers common safety and health hazards: lifting, cleaning solutions, slicing machines, protective clothing, housekeeping, electrical shock, and fires.

International Molders and Allied Workers Union (AFL-CIO)
Safety and Health Project
1216 East McMillan Street
Cincinnati, Ohio 45206
(513) 961-5141

Newsletter
Safety and Health Project Newsletter (2 pp., monthly)
Each issue contains several short feature articles on various hazards or on current events in the safety and health field. Simultaneously published in Spanish.

In Production (available August 1980)

Booklets
Safety Committees
Health and Safety Resources
How to Do a Walkaround Inspection
The Occupational Safety and Health Review Commission

Slide/Tape Show
"Hazards to Molders and Allied Workers"

Office and Professional Employees International Union (AFL-CIO)
Suite 610
265 West 14th Street

New York, New York 10011
(212) 675-3210

Booklet
Health Protection for Operators of VDTs/CRTs (16 pp., 1980, $.50)
Coproduced with the New York Committee for Occupational Safety and Health, this booklet describes video display terminals: how they work, what health problems they can cause, and what operators can do to reduce the hazards.

Oil, Chemical and Atomic Workers International Union (AFL-CIO)
Health and Safety Office
P.O. Box 2812
Denver, Colorado 80201
(303) 893-0811

Book
Peril on the Job (198 pp., 1970, $4.50 OCAW members/$5.50 nonmembers)
By Ray Davidson, this book reports, through the use of hundreds of workers' anecdotes, major hazards in the oil and chemical industries, relevant laws, and collective bargaining.

Booklets
Asbestos: Its Hazards and How to Fight Them (26 pp., 1978, $2/$3)
Written in a question-and-answer format, the booklet clearly explains asbestosis, health problems it causes, and what workers can do in case of exposure. Also contains some discussion of community exposure to asbestos. Includes OSHA asbestos standards.
Handbook for OCAW Women (40 pp., 1976, $1.25/$2)
Contains section on occupational safety and health for women workers.
Health Hazards in Petroleum Refineries (21 pp., 1973, $2/$3)
This publication is a set of flow sheets of the basic processes of oil refineries, together with hazards associated with each process. Processes include crude distillation, polymerization, phenol extraction, and the like.
Health Hazards of Nitroglycerin and Nitroglycol (11 pp., 1977, $2/$3)

Details health hazards of explosives manufactured: the effects of nitrates on the body. Also discusses OSHA standards and action workers can take to reduce exposure.

Hydrogen Sulfide (23 pp., 1977, $2/$3)
  Discusses hazards of exposure to hydrogen sulfide gas, protective ventilation and equipment, and OSHA standards. Includes a useful case study.

Oil Refinery Health and Safety Hazards (50 pp., 1975, $5/$15)
  Describes oil refinery operations and hazards; has an extensive section on OSHA, collective bargaining, and other methods the OCAW has used to reduce exposure to these hazards. Originally published by the Philadelphia Area Project on Occupational Safety and Health.

TDI: Is It Dangerous? (17 pp., 1976, $2/$3)
  Describes toluene diisocyanate (used in making urethane plastics): hazards, medical tests, handling procedures, ventilation, and other control strategies.

Factsheets
  Benzene—Cancer Risk Demands Cancer Controls (2 pp., 1977)
  Cancer in the Workplace—Part I (2 pp., 1975)
  Cancer in the Workplace—Part II (2 pp., 1976)
  Carbon Black, Graphite, and Calcinated Coke—Similar Industries, Similar Hazards (2 pp., 1977)
  Cooling It—The Problem of Heat in the Workplace (1 p.)
  Dye Manufacture—A Broad Spectrum of Occupational Hazards (2 pp., 1976)
  Ethylene Oxide Linked with Excess Leukemia Risk (1 p., 1977)
  Explosives—Dramatic and Insidious Hazards (2 pp., 1977)
  Hazards in the Corn Wet-Milling Industry (2 pp., 1977)
  Health Hazards in Paint Production (3 pp.)
  Health Hazards in Plastics Manufacture (2 pp.)
  Health Hazards in the Electronics Industry (4 pp.)
  Hydrogen Fluoride: Is It Dangerous? (4 pp., 1974)
  Job Exposures and Heart Disease (2 pp., 1979)
  Lung Disease—The Twentieth Century Plague (2 pp.)
  OCAW Women: Some Ideas for Action (4 pp., 1974)
  Welding: Unique Hazards Demand Special Controls (2 pp., 1976)

Newsletter
  Lifelines (4-6 pp., monthly, $5/year. Back issues available at $5/yr.)
    Highlights current events of interest to OCAW members in the occupational safety and health field. Frequently contains one- to two-page feature article on specific health or safety hazards.

Slide/Tape Show
"Asbestos: Fighting a Killer" (29 min., color, 1977, $125 sale)
Workers, medical experts, and union representatives describe the hazards of exposure to asbestos and detail the federal regulations workers can use to protect themselves from this dangerous substance. Accompanied by handbook that includes script, and the OCAW asbestos booklet.

Posters—Hazards Series
Asbestos (1979, $16.50/$26.50)
Hydrogen Sulfide ($18/$28)

Buttons and Bumper Stickers ($1 each)
"Job Safety and Health: A Right Not a Benefit"

International Brotherhood of Painters and Allied Trades
of the United States and Canada (AFL-CIO)
United Unions Building
1750 New York Avenue, N.W.
Washington, D.C. 20006
(202) 637-0700

Books
The IBPAT-OSHA Health and Safety Education Book (224 pp., 1978, $9.95)
    This comprehensive manual covers the health and safety hazards of painters and allied trades (drywall tapers, abrasive blasters, paint makers, carpet layers, glazers, floor tile and linoleum workers). Covers health and safety hazards including paint ingredients, falls, fires and explosions, compressed air and strains, as well as the Occupational Safety and Health Act and worker rights. Appendixes list common substances in the paint and allied trades, how to select personal protective equipment, and rights and responsibilities under the act.
The IBPAT-OSHA Health and Safety Education Book: Quiz Book (70 pp., 1978, $9.95)
    Designed for painters, abrasive blasters, tapers, paint makers, glazers, and floor coverers, the quiz book presents a series of questions on health and safety hazards, personal protective equipment, and safe work practices. To be used with videotape programs listed below.

Videotape Programs: 3/4" videotape cassettes (also available in 1/2", 1", and 2"), sound, color, all accompanied by instructor's script, student manual, and workbook.

	Purchase	Rental
General Programs (5-37 minutes)		
"Employee Rights and Responsibilities"	$75	$15
"About OSHA"	$60	$12.50
"No-Name Hospital"	$45	$10
"Accident Causation"	$75	$15
"When to Ask for Help: A Message from Dr. Edwin Holstein of the Mt. Sinai School of Medicine"	$60	$12.50
Specific Trade Programs (5-35 minutes)		
"Abrasive Blasters Health Hazards"	$60	$12.50
"Abrasive Blasters Safety Hazards"	$60	$12.50
"Bridge Scaffolding"	$60	$12.50
"Drywall Tapers Health Hazards"	$60	$12.50
"Drywall Tapers Safety Hazards"	$60	$12.50
"Floorcoverers Health Hazards"	$60	$12.50
"Floorcoverers Safety Hazards"	$60	$12.50
"General Health Hazards—Painters, Paint Makers, Abrasive Blasters or Floorcoverers"	$75	$15
"Introduction to IBPAT-OSHA Project"	$60	$12.50
"Manually Propelled Mobile Scaffolds"	$60	$12.50
"Paint Makers Safety Hazards"	$60	$12.50
"Paint Hazards"	$75	$15
Selected Respirator Programs (1-4 minutes)		
"Drywall Tapers Respirators"	$45	$10
"Floorcoverers Respirators—Isocyanate Vapors"	$45	$10
"General Painters Respirators—Bridge Painting"	$45	$10
"General Painters Respirators—Latex Painting"	$45	$10
"General Painters Respirators—Tank Interior Painting"	$45	$10
"Paint Makers Respirators—Organic Vapors"	$45	$10
"Paint Makers Respirators—Styrene Vapors"	$45	$10
"Painters/Floorcoverers Respirators—Bystander Exposure"	$45	$10
"Sandblasters Respirators—Carbon Monoxide"	$45	$10
"Sandblasters Respirators—Silica Sand/Bystander"	$45	$10

262 / SAFETY AND HEALTH CATALOGUE

	Purchase	Rental
Short or Spot Messages (1/2-5 minutes)		
"Death Plays a Hand"	$45	$10
"Falling Objects"	$45	$10
"Solvent Intoxication"	$45	$10
"Stop Contributing to Accidents"	$45	$10

United Paperworkers International Union (AFL-CIO)
163-03 Horace Harding Expressway
Flushing, New York 11365
(212) 762-6000

Pamphlet
Collective Bargaining on Occupational Health and Safety
(2 pp., 1976)
 Makes recommendations to UPIU local unions regarding structure and activities of safety and health committees.

Newspaper
The Paperworker (24 pp., monthly)
 Contains two monthly columns on job safety and health issues. "Your Work, Your Health" addresses specific health and safety problems faced by UPIU members, and "Safety and Health with OSHA" details recent regulatory and research events.

Operative Plasterers' and Cement Masons' International
 Association of the United States and Canada (AFL-CIO)
Safety and Health Program
1125 17th Street, N.W.
Washington, D.C. 20036
(202) 393-6576

Pamphlet
OSHA: The Law—Your Rights and Responsibilities (24 pp., 1979)
 Describes the Occupational Safety and Health Act, safety and health hazards, OSHA inspection procedures, and NIOSH.

In Production

Pamphlets
 Ladder Safety
 Scaffold Safety

Slide Shows
"Ladder Safety"
"Scaffold Safety"

Transparencies
"General Industry/Construction Safety and Health"

United Association of Journeymen and Apprentices of the
Plumbing and Pipefitting Industry of the United States
and Canada (AFL-CIO)
Department of Safety and Health
901 Massachusetts Avenue, N.W.
Washington, D.C. 20001
(202) 628-5823

Articles—Hazard Alerts
  Argon Gas Suffocates UA Members (2 pp., 1979)
    Describes the hazard of oxygen deficiency in the use of
    argon gas in confined spaces.
  Asbestos (3 pp.)
    Describes types of exposure to asbestos among plumbers and
    pipefitters, effects of asbestos on the body, and government
    standards.
  Contact Lenses on the Jobsite (2 pp., 1977)
    Discusses the relative safety of contact lenses and eyeglasses
    and offers tips on how to wear contact lenses safely.
  Members Exposed to Potential Explosion Hazard in Hospital
  Drains (1 p.)
    Describes explosion hazard caused by the formation of copper
    or lead azide in laboratory drains. Lists procedures for de-
    contaminating azide-containing plumbing systems.
  Plastic Solvent Cements and Primers (2 pp., 1977)
    Describes common plastic pipes and cements, hazards of
    exposure, and recommended handling procedures.
  Will We Have to Shout to Make Ourselves Heard on the Issue
  of Noise? (4 pp., 1976)
    Describes effects of noise, abatement methods, and govern-
    ment regulatory action.
  Work in Confined Spaces Poses Serious Hazards for UA Members
  (4 pp., 1977)
    Gives examples of confined space accidents and describes
    adequate respirator protection and other procedures to pre-
    vent these accidents.

Brotherhood of Railway, Airline and Steamship Clerks,
  Freight Handlers, Express and Station Employees (AFL-CIO)
  3 Research Place
  Rockville, Maryland 20850
  (301) 948-4910

Articles
  The Darker Side of Fluorescent Lighting (2 pp., 1976)
    Summarizes criticisms of fluorescent lighting: glare, eye
    strain, and excessive lighting levels.
  Grain Dust: Danger to Health and Safety (2 pp.)
    Reports on a BRAC-endorsed study of Canadian grain handlers.
  Grain Dust: Hazards Confirmed (1 p.)
    Announces the first report of the Canadian study.
  Video Display Terminals: A Look at Their Impact on Health and
  Safety (3 pp., 1979)
    Looks at some of the health hazards of VDT/CRTs and how to
    reduce or eliminate these hazards.

United Union of Roofers, Waterproofers and Allied Workers
  (AFL-CIO)
  Occupational Safety and Health Program
  1125 17th Street, N.W.
  Washington, D.C. 20036
  (202) 638-3228

Pamphlet
  Roofing Can Be a Dangerous Trade (2 pp., 1979)
    Introduces the Roofers' safety and health program.

United Rubber, Cork, Linoleum and Plastic Workers
  of America (AFL-CIO)
  Office of Industrial Hygiene
  87 South High Street
  Akron, Ohio 44308
  (216) 376-6181

Reports
  These reports summarize information about specific hazards to
  URW members.
    Antimony (7 pp., 1979)
    Asphalt Fumes (7 pp., 1979)
    Benzoyl Peroxide (12 pp., 1979)

Carbon Black (5 pp., 1979)
Chloroethanes (7 pp., 1979)
Diisocyanates (6 pp., 1979)
Epichlorohydrin (5 pp., 1979)
Fiberglass—Revised Comments Regarding the Effects of
  Employee Exposure (4 pp., 1979)
Fibrous Glass (11 pp., 1979)
Glycidyl Ether (15 pp., 1979)
Ketones (20 pp., 1979)
Nitrosamines in the Rubber Industry (6 pp., 1979)
Radiofrequency (RF) Sealers and Heaters (4 pp., 1980)
Refined Petroleum Solvents (7 pp., 1979)
Tetrachloroethylene (Perchloroethylene) (4 pp., 1979)
Vinyl Acetate (8 pp., 1979)
Vinyl Halides Carcinogenicity (4 pp., 1979)

Guidebook
  Rights and Resources: An Introductory Guide for URW Safety
  and Health Committees (17 pp., 1980)
    Introduces URW members to job safety and health, how to
    identify hazards and judge the degree of hazard, a guide
    to written information, how to document hazards and use
    the OSHAct. Also contains a list of organizations active in
    occupational safety and health.

URW 2980 Safety and Health Seminars (Instructor's Guides)
  Interpreting Industrial Hygiene Sampling Results Part I (4 pp.)
  Interpreting Industrial Hygiene Sampling Results Part II (5 pp.)
  OSHA Rights Workshop (3 pp.)
  Using the OSHA General Industry Standards 29 CFR 1910 (8 pp.)

Factsheets
  Dermatitis Is Frequent and Painful Ailment
  Electrical Hazards
  Headaches: More Than You Ever Wanted to Know
  Lifting: Minimizing Injury
  Noise: EPA Has Labeling Standard
  OSHA Ruling on Respirators
  Protection? (Protective Discrimination)
  Respirators
  Smoking in the Work Environment
  Some Scott Respirators May Need Modification
  Stress: There Is a Solution

Newsletter
: United Rubber Worker (12 pp., monthly, $1.20/year)
: : Frequently contains reports of URW locals' activities in health and safety and feature articles on specific hazards.

Service Employees International Union (AFL-CIO)
2020 K Street, N.W.
Washington, D.C. 20006
(202) 452-8750

Manuals
: OSHA and Bargaining in Health Care Facilities (19 pp., 1980)
: : Includes sections on preparation for bargaining, employee rights, negotiation, court decisions, and contract language.
: OSHA and Bargaining in the Building Service Industry (28 pp., 1977)
: : Includes sections on preparation for bargaining, employee rights, OSHA standards, negotiation, administrative and court decisions, contract clauses, how to file a complaint, and OSHA forms.

Booklet
: Hospital Safety and You (14 pp.)
: : Outlines common job hazards (for example, falls, strains) and what action workers can take to reduce or eliminate those hazards.

Newsletter
: Service Employee (12 pp., monthly)
: : Frequently contains articles on job safety and health issues.

Articles
: Burnout: The Professional Hazard You Face in Nursing (4 pp., 1978)
: : Discusses "burnout" syndrome in which overwork and stress can lead to emotional and physical exhaustion. Originally published in Nursing 78.
: Critique: Occupational Disease Among Operating Room Personnel (4 pp., 1975)
: : Challenges as lacking basis certain earlier studies that concluded that operating room personnel faced a hazard in exposure to waste anesthetic gases.
: Environmental Hazards in Hospitals (5 pp., 1974)

Summarizes common hospital hazards such as poor housekeeping, waste disposal, and infections spread through inadequate laundering of hospital bedding.

Guidelines to Assist Hospitals in the Use of Ethylene Oxide (3 pp., 1978)
 Originally published in Hospitals, Journal of the American Hospitals Association.

The Hospital: Hazardous Haven (6 pp.)
 Popular summary of major hospital hazards: back strain, falls, poor sanitation, infection, and inadequate housekeeping. Originally published in Job Safety and Health.

Hospital Hazards (10 pp., 1973)
 Gives brief descriptions of many potential hospital hazards and accidents. Originally published in The Practitioner.

Hospital Workers Face Hazards (2 pp., 1976)
 Describes a proposed occupational health and safety service for hospitals. Originally published in Occupational Health and Safety.

Management of Electrical Hazards in Hospitals (6 pp., 1972)
 Originally published in Canadian Hospital.

Occupational Health Hazards to Women: A Synoptic View (13 pp., 1977)
 Summarizes hazards to women workers. Sections on female-dominated occupations, hazards to pregnant women and fetuses, carcinogens, and other hazards peculiar to women.

Miscellaneous Materials
 The Research Department of SEIU has available additional articles on the Occupational Safety and Health Act and health care workers' occupational hazards.

American Federation of State, County and Municipal Employees (AFL-CIO)
1625 L Street, N.W.
Washington, D.C. 20036
(202) 452-4800

Booklet
Occupational Health and Safety of Municipal Workers (134 pp., 1977)
 Covers health hazards of sewage treatment sanitation, health care, and other municipal workers. Includes bibliography containing resource materials and organizations. Originally published by the Urban Environment Conference.

Report
   Survey of Public Sector Occupational Safety and Health Programs in Twelve States (29 pp., 1980)
      Evaluates sample state plan and nonstate plan states using the criteria of data collection, inspections, workplace monitoring, statewide committees, and self-evaluation. The report concludes that states with occupational safety and health state plans generally have better programs for employees of state and local governments than states without such programs.

American Federation of State, County and Municipal Employees (AFL-CIO), District Council 37
Occupational Safety and Health Project
140 Park Place
New York, New York 10007
(212) 766-1562

Booklets
   Health and Safety in the Office (10 pp.)
      Describes the physical, chemical, and stress-inducing hazards of office work. Originally published by Women Office Workers.
   Office Health and Safety Questionnaires (14 pp.)
   On the Road to Safety and Health: A Manual for Road Crews (75 pp., 1979, $1.50)
      Describes common safety and health hazards, and methods of elimination and control for highway repairers, traffic device maintenance personnel, and motor vehicle operators. Contains a hazard checklist.
   Workplace Health and Safety Survey (14 pp.)
      Compiles a series of questionnaires and survey forms for detecting a broad range of health and safety hazards.

Factsheets
   Asbestos (4 pp.)
   Bibliography—Occupational Safety and Health in Office Work (4 pp.)
   Caution: Office Zone (8 pp.)
   Checklist of Office Health and Safety Hazards (5 pp.)
   CRT/VDT Health Hazards: Questionnaire (4 pp.)
   Hazards for Clerical Workers (3 pp.)
   Heat Stress (5 pp.)
   How to Find Information About Industrial Materials (1 p.)

Noise (4 pp.)
Photocopiers (3 pp.)
Radiation, Microwaves and You (2 pp.)
Some General Information for Workers Regarding Safety and Health on the Job (4 pp.)
Varicose Veins (2 pp.)
What You Can Do (3 pp.)
What You Can Read (2 pp.)

Newsletter
  DC 37 Safety and Health News (12 pp., 3 issues/year)
    Covers health hazards of public-sector employees. Contains feature articles on specific job hazards and the hazards of specific occupations.

United Steelworkers of America (AFL-CIO)
Safety and Health Department
Five Gateway Center
Pittsburgh, Pennsylvania 15222
(412) 562-2581

Book
  Safety and Health Manual (190 pp.)
    Designed for local safety and health personnel, the manual summarizes suggested contract language, industrial hygiene and safety medical surveillance, and governmental agencies and standards.

Training Manual
  Hazard Recognition Program (106 pp.)
    Units: flammable liquids, welding, material handling, electricity, tools and protective equipment. Includes outline of hazards and standards and questions for student use. Also contains concise factsheets on these subjects.

Booklet
  Local Union Guidelines and Employee Rights Under the Occupational Safety and Health Act of 1970 (14 pp., 1978)
    Describes procedures for local unions, employee rights guaranteed in the OSHAct, and sample letters for a Freedom of Information Act request and for electing party status at the OSH Review Commission.
  Safety and Health Legislation (25 pp., 1979)
    Describes recent safety and health legislation.

Newsletter
  Safety and Health Update (6 pp., 4 issues/yr.)
    Reports on current activities of OSHA and the USWA Safety and Health Department. Contains feature articles on job hazards and abatement techniques.

Speciality Conferences
  The USWA Safety and Health Department conducts conferences on these topics:
    Lead
    Aluminum
    Chemical
    Steel
    Foundry
    Employee Rights and Review Commission
  Conferences cover health and safety hazards in the particular industry, including relevant OSHA standards. Conferences are accompanied by handout materials, and most include audio-visual materials.

Reporting Forms
  Investigation Report (1 p.)
  Safety and/or Health Inspection Report (1 p.)
  Safety or Health Recommendations (1 p.)

Films
  "The American Way of Cancer" (50 min., color, 1975)
    This film places strong emphasis on research linking cancer to the chemicals in the air we breathe, the food we eat, the water we drink, and the toxic substances we are exposed to at work. The film reminds us that we know more and more about the causes of cancer and must now move to control these toxic substances and reduce exposure to the lowest possible level. Produced by CBS.
  "Foundry Safety and Health Test" (28 min., color)
    Presents 19 hazardous situations in foundry operations and asks viewers to answer questions about the hazards. Presents solutions using scenes from actual foundry operations.
  "Health Hazard in the Shop" (25 min., color, 1978)
    The film is intended to familiarize viewers with industrial hygiene inspection procedures, equipment used, and rights and responsibilities of employee representatives and management when a health hazard is suspected. Produced by the University of Wisconsin Extension School for Workers.

"Listen" (45 min., color, 1974)
   This film tells the story of an older worker who has long been exposed to damaging noise levels in a paper mill with no understanding that the disability he now suffers is work-related. This film can help build recognition of on-the-job health hazards and how OSHA can be used to reduce or eliminate them. Produced by the United Paperworkers International Union.
"More Than a Paycheck" (28 min., color)
   Explores the cancer risk faced by industrial workers. The film recommends modified engineering and work practices for the control of exposure to carcinogenic substances. Produced by George Washington University, under contract with the Occupational Safety and Health Administration.
"Shop Accident" (25 min., color, 1976)
   The film covers: right to file a complaint anonymously, right to walk around with the inspector, how safety hazards are identified, compliance officer discussing findings with management and the union committee. Produced by the University of Wisconsin, Extension School for Workers.
"Working for Your Life" (57 min., color, 1979)
   The film focuses on the hazards faced by today's 43 million American working women. It is the only documentary film about the health and safety of women on the job. Produced by the Labor Occupational Health Program.

International Woodworkers of America (AFL-CIO)
1622 North Lombard Street
Portland, Oregon 97217
(503) 285-5281

Factsheets (1-4 pp.)
   Cedar Dust Asthma Study Completed
   Chain Saw Vibration
   Does Your Chain Saw Measure Up?
   Economics: The Human Factor in Safety
   The Great Unknown: Are We Guinea Pigs?
   Noise Control Working at Delta Plywood
   Occupational Safety and Health
   Satisfaction Sought in Swedish Sawmills
   Studies Show Power Saws Cause "White Finger"
   Why Do Accidents Happen?
   Wood Preservatives—Hazards and Precaution

Newsletter

<u>International Woodworkers</u> (8 pp., monthly)
    Contains occasional articles on new developments and union policies on occupational safety and health.

Film

"Safety and Health in the Woodworking Industry" (20 min., color)
    The film follows the lumber industry process from forest to milling, noting common safety and health hazards along the way. The film includes a discussion of hazard remedies.

SUBJECT INDEX

General

General Works

  <u>ACTWU Local Union Health and Safety Action Manual</u>
  <u>AIW Safety and Health Training Manual</u>
  <u>Health and Safety Manual</u>, ICWU
  <u>How to Put the Occupational Safety and Health Act to Work: A Guideline for Stewards</u>, Paperworkers
  <u>IAM Guide for Safety and Health Committees</u>
  <u>OSHA: The Law—Your Rights and Responsibilities</u>, Plasterers
  <u>Safety and Health Manual</u>, USWA
  Some General Information for Workers Regarding Safety and Health on the Job, DC 37
  <u>Watch Out! A Handbook of Safety and Health for Furniture Workers</u>, UFWA
  <u>What Every UAW Representative Should Know About Health and Safety</u>
  <u>Work Without Fear</u>, IUD

Hazard Identification and Abatement

  "Accident Causation," IBPAT
  ACTWU Checklist of Local Union Actions in Connection with OSHA Complaints and Inspections
  ACTWU Checklist on Hazard Correction
  ACTWU Questionnaire for Hazard Investigation
  <u>Checklist of Possible Health and Safety Problems in Your Shop</u>, District 65
  Controlling Deadly Vapors, UAW
  Cracking Codes on Company Chemicals, ACTWU

CRT/VDT Health Hazards: Questionnaire, DC 37
Design of Local Exhaust Ventilations, ACTWU
Health Hazard Inventory, ICWU
Health Hazard Questionnaire, ICWU
How to Crack the Company's Code: Getting Names of Workplace Chemicals, UAW
How to Do a Walkaround Inspection, Molders
How to Evaluate On-The-Job Health Hazards, UAW
How to Find Information about Industrial Materials, DC 37
"I Never Had an Accident in My . . .," Meat Cutters/Retail Food Store Employees Local 342
Interpreting Industrial Hygiene Sampling Results, Part I, URW
Interpreting Industrial Hygiene Sampling Results, Part II, URW
Lead: A Worker's Guide to Checking Exposure to Lead, UAW
Machinery Lockout Procedures, UAW
"Stop Contributing to Accidents," IBPAT
Unsafe Acts as a Cause of Accidents, ACTWU
Ventilation, UAW
Why Do Accidents Happen? IWA
Work Without Fear, IUD
Workplace Health and Safety Survey, DC 37

Medical Tests and Information

Audiometric Monitoring (Hearing Tests) and the Meaning of Test Results, ACTWU
Demand for Bargaining and Request for Company Information—Medical Examinations, ICWU
Glossary of Health Hazard Terms, ACTWU
Help for the Working Wounded, IAM
Job Exposures and Heart Disease, OCAW
Lung Disease—The Twentieth Century Plague, OCAW
Lung Function Tests, UAW
"When to Ask for Help," IBPAT

OSHA—Guides to Standards and Procedures

"About OSHA," IBPAT
Barlow Decision, IAM
Bureau of Motor Carrier Safety Regulations on Employee Safety and Health in the Operation, Maintenance and Loading and Unloading of Motor Vehicles, IAM
"Employee Rights and Responsibilities," IBPAT
Federal Executive Order 12196, IAM
Guidebook to Occupational Safety and Health, AFL-CIO Building and Construction Trades Department

274 / SAFETY AND HEALTH CATALOGUE

"Health Hazard in the Shop," USWA
How to Get Information from OSHA, UAW
How to Use OSHA Form 200, ACTWU
How to Use the OSHA Standards, ACTWU
The Occupational Safety and Health Review Commission, Molders
OSHA, UAW
OSHA Forms, ACTWU
    Material Safety Data Sheet
    Supplementary Record of Occupational Injuries and Illnesses
<u>OSHA General Industry Standards</u>, ICWU
OSHA Guidelines on Hearing Conservation Program, ACTWU
OSHA Inspection Check-List, ICWU
OSHA Lead Standard—Summary of Medical Surveillance and
   Job-Transfer Sections, ACTWU
OSHA Respirator Regulations Guide, ACTWU
OSHA's Asbestos Standard Guide, ACTWU
OSHA: The Law—Your Rights and Responsibilities, Plasterers
<u>Safety and Health Legislation</u>, USWA
"The Shop Accident," USWA
Status of State Plans, IAM
<u>Using the OSHA General Industry Standards 29 CFR 1910</u>, URW
Walk-Around Violations, IAM
Whirlpool Decision, IAM

Personal Protective Equipment

  See International Association of Fire Fighters
"Drywall Tapers Respirators," IBPAT
Earplugs, ACTWU
"Floorcoverers Respirators—Isocyanate Vapors," IBPAT
"General Painters Respirators—Bridge Painting," IBPAT
"General Painters Respirators—Latex Painting," IBPAT
"General Painters Respirators—Tank Interior Painting," IBPAT
Health Hazards and Personal Protective Equipment, AFL-CIO
   Building and Construction Trades Department
OSHA Respirator Regulations Guide, ACTWU
OSHA Ruling on Respirators, URW
"Paint Makers Respirators—Organic Vapors," IBPAT
"Paint Makers Respirators—Styrene Vapors," IBPAT
"Painters/Floorcoverers Respirators—Bystander Exposure,"
   IBPAT
Protective Equipment, USWA
Respirators, URW
Respiratory Protection Devices, ICWU
"Sandblasters Respirators—Carbon Monoxide," IBPAT

APPENDIX I / 275

"Sandblasters Respirators—Silica Sand/Bystander," IBPAT
Some Scott Respirators May Need Modification, URW
Strapped to a Mask? Here Are the Rules for Respirators, UAW

Resource Guides

Bibliography—Occupational Safety and Health in Office Work, DC 37
Education Materials, ACTWU
Health and Safety Resources, Molders
What You Can Read, DC 37

Sample Forms and Letters

Closing Conference Demand (OSHA), ICWU
Complaint Against State Program Administration, ICWU
Contest of Abatement Date (Review Commission), ICWU
Demand for Bargaining and Request for Company Information—Absenteeism, ICWU
Demand for Bargaining and Request for Company Information—Female Exclusion Policy, ICWU
Demand for Bargaining and Request for Company Information—Medical Examinations, ICWU
11(c) Discrimination Complaint (OSHA), ICWU
Freedom of Information Act Request, ICWU
How to Use the OSHA Form 200, ACTWU
Imminent Danger (OSHA), ICWU
Inspection Request (OSHA), ICWU
Investigation Report, USWA
OSHA Forms, ACTWU
    Material Safety Data Sheet
    Supplementary Record of Occupational Injuries and Illnesses
Request for Company Information—General Health and Safety, ICWU
Request for Party Status (Review Commission), ICWU
Safety and/or Health Inspection Report, USWA
Safety or Health Recommendations, USWA

Union Activities and Contract Language

Action Planning, GAIU
ACTWU Actions to Protect Worker Safety and Health
ACTWU Checklist of Local Union Actions in Connection with OSHA Complaints and Inspections
ACTWU Checklist on Hazard Correction
Collective Bargaining on Occupational Health and Safety, Paperworkers

276 / SAFETY AND HEALTH CATALOGUE

Combating Hazards on the Job: A Workers' Guide, AFL-CIO, Food & Beverage Trades Dept.
Forming CWA Local Occupational Safety and Health Committees
General Policy of the ICWU on Matters of Health and Safety
Health and Safety Clause, UAW District 65
How to Crack the Company's Code: Getting Names of Workplace Chemicals, UAW
How to Use the Lead Standard: A Local Union Action Program, UAW
IAM Guide for Safety and Health Committees
ICWU Policy on Reproductive Effects of Hazardous Materials
Local Union Guidelines and Employee Rights Under the Occupational Safety and Health Act of 1970, USWA
Local Union Health and Safety Records, ACTWU
OSHA and Bargaining in Health Care Facilities, SEIU
OSHA and Bargaining in the Building Service Industry, SEIU
OSHA Inspection Check-List, ICWU
OSHA Rights Workshop, URW
Pacific Coast Longshore Contract Document 1978-81, ILWU
Rights and Resources: An Introductory Guide for URW Safety and Health Committees
Safety Committee, Molders
The Safety Committee and Collective Bargaining: New Concepts, Paperworkers
Safety Committee Request Guidelines, ACTWU
Suggested Health and Safety Language, ICWU
Watch Out! A Handbook of Safety and Health for Furniture Workers, United Furniture Workers

Women's Occupational Safety and Health

Demand for Bargaining and Request for Company Information—Female Exclusion Policy, ICWU
Handbook for OCAW Women
OCAW Women: Some Ideas for Action
Occupational Health Hazards to Women: A Synoptic View, SEIU
Protection? (Protective Discrimination), URW
Women Workers: Hazards on the Job, AFL-CIO Department of Occupational Safety and Health
"Working for Your Life," USWA

Workers' Compensation

Workers' Compensation, ACTWU
Workers' Compensation Facts for Synthetic Division Members, ACTWU

Hazards

Accidents and Accident Causation

"Accident Causation," IBPAT
"I Never Had an Accident in My . . .," Meat Cutters/Retail Food Store Employees Local 342
"Stop Contributing to Accidents," IBPAT
Unsafe Acts as a Cause of Accidents, ACTWU
Why Do Accidents Happen? IWA

Asbestos

Asbestos, CWA
Asbestos, DC 37
Asbestos, IAM
Asbestos, UA
Asbestos (Poster), OCAW
Asbestos and Pipefitters, ACTWU
"Asbestos: Fighting a Killer," OCAW
Asbestos: Its Hazards and How to Fight Them, OCAW
The Law: What It Demands in the Process of Stripping Asbestos, Asbestos Workers
OSHA's Asbestos Standard—Guide, ACTWU

Chemical

"The American Way of Cancer," USWA
Antimony, URW
Arsenic, CWA
Asphalt Fumes, URW
Benzene, IAM
Benzene—Cancer Risk Demands Cancer Controls, OCAW
Benzoyl Peroxide, URW
Beryllium, IAM
Cancer in the Workplace—Part I, OCAW
Cancer in the Workplace—Part II, OCAW
Carbon Black, URW
Carbon Black, Graphite and Calcinated Coke—Similar Industries, Similar Hazards, OCAW
Carbon Monoxide, CWA
Chloroethanes, URW
Chloroprene, URW
Common Toxic Substances and Their Effects, UAW
Controlling Deadly Vapors, UAW
Cracking Codes on Company Chemicals, ACTWU
Cutting Oils, IUE Local 201

278 / SAFETY AND HEALTH CATALOGUE

Diisocyanates, URW
Dye Manufacture—A Broad Spectrum of Occupational Hazards, OCAW
Epichlorohydrin, URW
Ethylene Oxide Linked with Excess Leukemia Risk, OCAW
Explosives—Dramatic and Insidious Hazards, OCAW
<u>Fiberglass—Revised Comments Regarding Health Effects of Employee Exposure</u>, URW
<u>Fibrous Glass</u>, URW
Fluorides, ICWU
Freon, CWA
<u>Glycidyl Ether</u>, URW
Hazardous Substances, UAW
Hazards in the Corn Wet-Milling Industry, OCAW
Health Hazards Associated with Diesel Engine Emissions, IAM
Health Hazards in Coal Tar Production, OCAW
<u>Health Hazards in Petroleum Refineries</u>, OCAW
Health Hazards in Paint Production, OCAW
Health Hazards in Plastics Manufacture, OCAW
Health Hazards in the Electronics Industry, OCAW
<u>Health Hazards of Nitroglycerin and Nitroglycol</u>, OCAW
<u>Help for the Working Wounded</u>, IAM
How to Crack the Company's Code: Getting Names of Workplace Chemicals, UAW
How to Prevent Skin Disease Caused by Cutting Fluids, UAW
Hydrogen Fluoride: Is It Dangerous? OCAW
<u>Hydrogen Sulfide</u>, OCAW
Hydrogen Sulfide (Poster), OCAW
Industrial Solvents, IAM
Job Exposures and Heart Disease, OCAW
<u>Ketones</u>, URW
Lung Disease—The Twentieth Century Plague, OCAW
"More Than a Paycheck," AFL-CIO Food and Beverage Trades Department; IAM; USWA
<u>Nitrosamines in the Rubber Industry</u>, URW
<u>Oil Refinery Health and Safety Hazards</u>, OCAW
<u>Peril on the Job</u>, OCAW
Pesticides, ICWU
Plastic Solvent Cements and Primers, UA
Polychlorinated Biphenyls, CWA
Polyurethane, ICWU
Polyurethane: Job Health Hazard? UAW
Polyurethanes and Isocyanates, CWA
<u>Refined Petroleum Solvents</u>, URW
Smoking in the Work Environment, URW

APPENDIX I / 279

Solvent Intoxication, IBPAT
TDI: Is It Dangerous? OCAW
Tetrachloroethylene (Perchloroethylene), URW
Toxic Chemicals, GAIU
Vinyl Acetate, URW
Vinyl Chloride, CWA
Vinyl Halides Carcinogenicity, URW
Wood Preservatives—Hazards and Precaution, IWA

Dusts

Carbon Black, Graphite and Calcinated Coke—Similar Industries, Similar Hazards, OCAW
Cedar Dust Asthma Study Completed, OWA
Cotton Dust Control Manual, ACTWU
Fiberglass—Revised Comments Regarding Health Effects of Employee Exposure, URW
Fibrous Glass, URW
Grain Dust: Danger to Health and Safety, BRAC
Grain Dust: Hazards Confirmed, BRAC
Hazards in the Corn Wet-Milling Industry, OCAW
"Our Lives, Our Rights," ACTWU

Fires and Explosions

Bearing Hazards, AFL-CIO Food and Beverage Trades Department
Bucket Elevator Hazards and Design, AFL-CIO Food and Beverage Trades Department
Conveyor Hazards and Design, AFL-CIO Food and Beverage Trades Department
Distribution Floor and Bin Design, AFL-CIO Food and Beverage Trades Department
Dust Collection Design and Operation, AFL-CIO Food and Beverage Trades Department
Electrical Equipment Standards, AFL-CIO Food and Beverage Trades Department
Explosion Suppression and Inerting, AFL-CIO Food and Beverage Trades Department
Explosion Venting, AFL-CIO Food and Beverage Trades Department
Explosives—Dramatic and Insidious Hazards, OCAW
Fire Brigades, ACTWU
Flammable Liquids, USWA
General Fire Prevention and Emergency Procedures, AFL-CIO Food and Beverage Trades Department

280 / SAFETY AND HEALTH CATALOGUE

    Members Exposed to Potential Explosion Hazard in Hospital Drains, UA
    Static Electricity Removal, AFL-CIO Food and Beverage Trades Department

Lead

    <u>Control of Lead in Battery Making</u>, UAW
    <u>The Hazards of Lead and How to Control Them</u>, UAW
    How to Use the Lead Standard: A Local Union Action Program, UAW
    Lead, CWA
    Lead, IAM
    Lead: A Worker's Guide to Checking Exposures to Lead, UAW
    <u>Lead Hazards in Vehicle Assembly and Repair Operations: Body Solder</u>, UAW
    OSHA Lead Standard—Summary of Medical Surveillance and Job Transfer Sections, ACTWU

Noise

    Audiometric Monitoring (Hearing Tests) and the Meaning of Test Results, ACTWU
    "Listen," USWA
    Noise, DC 37
    Noise, GAIU
    Noise, IAM
    Noise, UAW
    Noise Case Study, ACTWU
    <u>Noise Control: A Worker's Manual</u>, UAW
    Noise Control Working at Delta Plywood, IWA
    Noise: EPA Has Labeling Standard, URW
    <u>Noise Reduction in Printing Plants</u>, GAIU
    OSHA Guidelines on Hearing Conservation Program, ACTWU
    Will We Have to Shout to Make Ourselves Heard on the Issue of Noise? UA

Physical and Safety

    Argon Gas Suffocates UA Members, UA
    Beat the Heat, ICWU
    Chain Saw Vibration, IWA
    Contact Lenses on the Jobsite, UA
    Cooling It—The Problem of Heat in the Workplace, OCAW
    CRT/VDT Health Hazards: Questionnaire, AFSCME DC 37
    The Darker Side of Fluorescent Lighting, BRAC
    Electrical Hazard Control, AFL-CIO Building and Construction Trades Department

APPENDIX I / 281

Electrical Hazards, URW
Electrical Hazards, Fires, Walking and Working Surfaces, GAIU
Electricity, USWA
Ergonomics, GAIU
"Falling Objects," IBPAT
Guide to Safe Rigging Practices, Boilermakers 169
Health Protection for Operators of VDTs/CRTs, OPEIU
Health Problems from Lifting and Carrying, Boilermakers Local 802
Heat Stress, DC 37
Job Stress, GAIU
Lifting: Minimizing Injury, URW
Material Handling, USWA
Radiation, IAM
Radiation—Nonionizing, IAM
Radiation and Radiation Protection, Boilermakers Local 169
Radiation, Microwaves and You, DC 37
Radiofrequency (RF) Sealers and Heaters, URW
Shift Work, ICWU
Stress: There is a Solution, URW
Studies Show Power Saws Cause "White Finger," IWA
Video Display Terminals, CWA
Video Display Terminals: A Look at Their Impact on Health and Safety, BRAC
Work in Confined Spaces Poses Serious Hazards for UA Members

Occupations

Construction

All AFL-CIO Building and Construction Trades Department Programs
All IBPAT Programs
All UA Programs
"General Industry/Construction Safety and Health," Plasterers
Guide to Safe Rigging Practices, Boilermakers 169
Ladder Safety, Plasterers
Scaffold Safety, Plasterers

Foundry

See International Molders and Allied Workers Union
Foundry, USWA
Foundry Fatality Prevention, UAW
Foundry Safety and Health Test, USWA
UE Manual on Foundry Health and Safety

282 / SAFETY AND HEALTH CATALOGUE

Grain

See AFL-CIO Food and Beverage Trades Department
Hazards in the Corn Wet-Milling Industry, OCAW
Grain Dust: Hazard to Health and Safety, BRAC
Grain Dust: Hazards Confirmed, BRAC

Hospital and Health Care

See Service Employees International Union
Members Exposed to Potential Explosion Hazard in Hospital Drains, UA
Varicose Veins, DC 37

Office

See AFSCME District Council 37
Health Protection for Operators of VDTs/CRTs, OPEIU

Public Sector

Federal Executive Order 12196, IAM
Occupational Health and Safety of Municipal Workers, AFSCME
Occupational Safety and Health: A Promise Unfulfilled for Public Employees, AFL-CIO Public Employee Department
On the Road to Safety and Health: A Manual for Road Crews, DC 37
Survey of Public Sector Occupational Safety and Health Programs in Twelve States, AFSCME

Shipyard

See International Brotherhood of Boilermakers
See International Longshoremen's and Warehousemen's Union
Safe Staging on Ships, Boilermakers Local 802

Welding

Precautions and Safe Practices for Boilermaker Welders, Boilermakers Local 169
Welding, IUE Local 201
Welding, OCAW
Welding, USWA
Welding: Unique Hazards Demand Special Controls, OCAW

OTHER INDEXES

Newsletters

Safety and Health

<u>DC 37 Safety and Health News</u>
<u>Health and Safety Bulletin</u>, IUE
<u>IAM Safety Gram</u>
<u>IUD Facts and Analysis</u>
<u>Lifelines</u>, OCAW
<u>Occupational Health and Safety Newsletter</u>, UAW
<u>Safety and Health Project Newsletter</u>, Molders
<u>Safety and Health Update</u>, USWA

General

<u>ACTWU Labor Unity</u>
<u>BC&T News</u>
<u>BC&T Report</u>
<u>The Chemical Worker</u>, ICWU
<u>The International Fire Fighter</u>, IAFF
<u>International Woodworkers</u>, IWA
<u>The Paperworker</u>, UPIU
<u>United Rubber Worker</u>, URW
<u>Service Employee</u>, SEIU
<u>UE News</u>

Audiovisual Materials

Films and Videotapes

See International Brotherhood of Painters and Allied Trades of the United States and Canada
"The American Way of Cancer," USWA
"Foundry Safety and Health Test," USWA
"General Industry/Construction Safety and Health," Plasterers
"Health and Safety in the Woodworking Industry," IWA
"Health Hazard in the Shop," USWA
"I Never Had an Accident in My . . .," Meat Cutters/Retail Food Store Employees Local 342
"Listen," USWA
"More Than a Paycheck," AFL-CIO Food and Beverage Trades Department; IAM; USWA
"Paul Jacobs and the Nuclear Gang," IAM
"The Shop Accident," USWA
"Working for Your Life," USWA

Slide/Tape Shows

"Asbestos: Fighting a Killer," OCAW
"Hazards to Molders and Allied Workers," Molders
"Ladder Safety," Plasterers
"Local 342 Safety Training Program," Meat Cutters/Retail Food Store Employees Local 342
"Our Lives, Our Rights," ACTWU
"Scaffold Safety," Plasterers

Foreign Language Materials

Spanish (all UAW unless otherwise noted)

Safety and Health Project Newsletter, Molders
Arsénico
Asbestos
Como Prevenir Enfermedades de la Piel Causadas por los Fluidos de Cortar
Dermatitis
Diseño para una Ventilación de Extracción Local
El Effecto de Calor
¿Está Su Cara Liada a una Mascara? Estas Son las Reglas que Rigen a los Respiradores
Fibras de Vidrio
Fluorocarburos
El Formaldehido
Lo Que Todo UAW Representante Debe Saber Sobre la Salud y la Seguridad
Monoxido de Carbono
Los Peligros que Represente El Plomo y Como Controlarlos
Silicosis
Tricloroetileneo
Ventilación

Portuguese

Cloro, UAW
Dermatite, UAW

French

La Prevention Des Accidents Mortels Dans Les Fonderies, UAW

APPENDIX J: INTERNATIONAL ORGANIZATIONS

WORLD HEALTH ORGANIZATION

On July 22, 1946, 61 nations of the International Health Conference met in New York and signed the constitution of the World Health Organization (WHO), and the WHO Interim Commission was established. It was ratified by 26 nations on April 7, 1948.

World Health Organization, which has its headquarters in Geneva, Switzerland, forms part of the United Nations system. It is a specialized agency, provided for in the United Nations, which has a formal agreement with the United Nations that provides, inter alia, for reciprocity between the two organizations, the exchange of information, and the adoption of common administrative practices. There are formal agreements between WHO and the Pan American Health Organization (PAHO), the International Labor Organization (ILO), the Food and Agriculture Organization (FAO), the United Nations Educational, Scientific and Cultural Organization (UNESCO), and the International Atomic Energy Agency (IAEA).

The constitution of WHO defines the organization's objective as "The attainment by all peoples of the highest level of health."

Occupational Health Program

WHO works in close collaboration with the International Labor Organization, which is concerned not only with occupational health but also with the wider field of the economic and social well-being of the worker. A priority area in this activity is assistance, on request, to developing countries to help them in forseeing and preventing hazards that can arise from the introduction of industrialization and stimulation of research on newer occupational health problems.

WHO has advocated a strong role for public health in the occupational health field, particularly in developing countries where the work environment in such areas as agriculture and animal husbandry is inseparable from the total health picture. WHO believes that the emphasis on occupational health should be more hazard-prevention oriented to include preplacement examinations, immunizations, nutrition, early diagnosis, and treatment of all kinds of diseases and health education.

Closely tied to the occupational health issues of hazard prevention and worker well-being is the issue of occupational health services. WHO sees national centralization as necessary to establish medical and health standards for licensing new workplaces, for compiling statistics, for supervising local services, and for training. WHO's emphasis is on health problems and health care, rather than on inspection and, therefore, it promotes the health authority rather than the labor authority as the coordinator of occupational health programs.

The program activities of WHO are geared to assist in hazard prevention, worker well-being, and national occupational health services. WHO provides direct assistance to member states by helping to establish occupational health institutes, awarding fellowships, conducting seminars and training courses, supporting research efforts, and providing consultants. The organization has contributed to international research on occupational health information dissemination and training efforts, and it has encouraged each member country to exert the same research and training efforts.

Environmental Health Program

WHO's activities are directed toward promoting and advising the international community on the prevention of water-borne and food-borne diseases by the provision of safe and adequate water supplies, the sanitary collection and disposal of human wastes, and the hygienic processing and distribution of foods.

WHO advises many governments on national programs for the provision of basic sanitary services and, especially, on the best way of using locally available material and manpower resources. It also assists with the training of professional and auxiliary personnel and provides guides on the planning, design, construction, maintenance, and operation of services for water supply and waste disposal, hygienic housing, and the hygiene of food-handling establishments. The WHO publication <u>International Standards for Drinking-Water</u> reached its third edition by 1971.

The major obstacle to the development of community services for adequate water supplies and the sanitary disposal of wastes is insufficient financing. To help in overcoming this obstacle, WHO organized, in collaboration with the World Bank, a preinvestment planning program that draws on the knowledge and experience of engineers, economists, financial analysts, and management experts. The main object of this program is to assist governments to elaborate firm plans developed under the United Nations Development Program for constructing plants and institutions that will be considered by external sources of financing—notably the World Bank, the Regional Development Banks, and bilateral sources—as worthy of support. By 1974 more than 40 countries had been assisted in this way. Projects completed in seven countries represented a capital value of almost $400 million.

While in the developing countries the environmental pollutants of major importance are still of biological origin, the industrialized countries are exposed to an increasing variety and concentration of chemical and physical pollutants of both air and water. The growing international concern with this problem was reflected in the holding in 1972 of the United Nations Conference on the Human Environment, in which WHO actively participated. That carefully planned antipollution measures can be of extraordinary effectiveness is illustrated by the experience of the capital city of a European country that enacted special clean air legislation in 1956. In less than two decades the annual amount of sunshine recorded had doubled. Moreover, thanks to restrictions on water pollution, the stretch of river running through the city became populated with fish for the first time in living memory.

WHO assists governments in the planning, operation, and evaluation of services for the control of environmental pollution. This involves the qualitative and quantitative assessment of sources of pollution, the establishment of criteria for maximum acceptable concentrations, and the development of methods for satisfying such criteria. WHO also systematically collects and evaluates information on the effects on health of pollutants in air, water, food, and the working environment.

ADDRESSES OF WHO HEADQUARTERS, REGIONAL OFFICES,
THE INTERNATIONAL AGENCY FOR RESEARCH ON CANCER,
AND THE WHO LIAISON OFFICE WITH THE UNITED NATIONS

World Health Organization

Headquarters
World Health Organization
1211 Geneva 27
Switzerland

Africa
World Health Organization,
Regional Office for Africa
P.O. Box 6
Brazzaville, Congo

The Americas
World Health Organization,
Resional Office for the Americas/Pan American Sanitary Bureau
525 23rd Street, N.W.
Washington, D.C. 20037

Eastern Mediterranean
World Health Organization,
Regional Office for the Eastern Mediterranean
P.O. Box 1517
Alexandria, Egypt

Europe
World Health Organization,
Regional Office for Europe
8, Scherfigsvej
2100 Copenhagen O, Denmark

South-East Asia
World Health Organization,
Regional Office of South-East Asia
World Health House, Indraprastha Estate
Ring Road
New Delhi-1, India

Western Pacific
World Health Organization,
Regional Office for the Western Pacific
P.O. Box 2932
12115 Manila, Philippines

International Agency for Research on Cancer (IARC)
150 Cours Albert Thomas
69372 Lyon, France

Contact: Dr. L. Tomatis, Chief, Unit of Chemical Carcinogenesis

The IARC conducts research and programs on the evaluation of the carcinogenic risk of chemicals to humans. It provides monographs on The Evaluation of the Carcinogenic Risk of Chemical to Humans, Volumes 1-20, World Health Organization, Geneva, 1972

Liaison Office with the United Nations
World Health Organization
New York, New York 10017

INTERNATIONAL LABOR ORGANIZATION

The International Labor Organization (ILO) was created by the Peace Treaty of Versailles in 1919 alongside the League of Nations, of which it was an autonomous part to promote social progress, without which there can be no harmonious economic or social development. The 140 member states of the ILO subscribe to the principles written into its constitution. They cooperate in its work, which they also finance. They are represented at all levels of the organization by government and workers' and employers' delegates who confer together on a basis of equality.

The ILO helps work out development policies, and it strives to ensure that the fundamental rights of workers are protected. The unique feature of the ILO is its tripartite structure. Its governing body is composed of government, employer, and worker members; a general conference of representatives of the member nations; and an International Labor Office controlled by the governing body.

The ILO has three major activities: standard setting, research, and technical assistance—it has been particularly active in the area of standard setting. It has adopted 140 conventions

---

4, Chemin des Morillons
1211 Geneva 22 Switzerland
Telephone: 99.67.16; Telex: 22.271

that, when ratified by a nation, oblige that government to bring its laws into conformity with them. The ILO has also adopted 148 recommendations that serve as guidelines for national action. Nearly one-third of the conventions and recommendations apply to occupational health and safety.

In 1975 the ILO adopted a convention and recommendation on occupational cancer and the ILO is presently considering a convention to protect workers against workplace hazards due to air pollution, noise, and vibration.

Exchange of scientific and technical information as an ILO goal is implemented through symposia, conferences, and courses, which are often jointly sponsored with WHO and other international bodies. Past symposia topics have included radiation protection, ergonomics and machine design, safety of mobile electric tools and appliances, ergonomics and environmental factors, and prefabricated building safety.

In addition to a number of standard regulation booklets, guides and handbooks, the ILO has published a major reference work, the Encyclopaedia of Occupational Health and Safety, which brings together contributions by 700 authors in 70 countries in two volumes. The Encyclopaedia deals with all aspects of occupational accident prevention, the improvement of hygiene at work, and the protection of the workers' health. In addition to the Encyclopaedia, model codes, codes of practice, and other technical manuals, the ILO publishes other materials, including an occupational safety and health series on toxic substances, ergonomics, dust prevention, radiation, and other topics; a film collection; and reports on meetings, symposia, and conferences. The most extensive effort of the ILO in information collection and dissemination is the International Occupational Safety and Health Information Center [CIS]. Begun in 1919 as a voluntary cooperative effort by the ILO and other national and international bodies, the CIS has been collecting, analyzing, and disseminating information and the results of experience gained throughout the world. The CIS is supported in this work by 34 national centers.

The ILO has been providing various forms of technical assistance since its beginning in 1919. ILO experts help national administrations create or strengthen their services. The ILO sends teams of experts for a period of six months to one year to study a specific country's problems. The countries thus served are able to modernize their laws and regulations, develop more advanced techniques for hazard control, or set up special inspectorates, training programs, and research laboratories. Over 50 countries have been involved in some form of ILO technical assistance.

APPENDIX J / 291

Bibliography

Ashford, Nicholas A., George I. Heaton, Judith I. Katz, and Sally T. Owen. "The Foreign Experience and Its Relevance to the United States." In <u>Protecting People at Work: A Reader in Occupational Safety and Health</u>. Washington, D.C.: U.S. Department of Labor, 1980.

International Labor Office. <u>The ILO and the World at Work</u>. Geneva, July 1979.

World Health Organization. <u>Introducing WHO</u>. Geneva, 1976.

ILO Member States

Afghanistan	Cuba
Algeria	Cyprus
Angola	Czechoslovakia
Argentina	Democratic Kampuchea
Australia	Democratic Yemen
Austria	Denmark
Bahamas	Djibouti
Bahrain	Dominican Republic
Bangladesh	Ecuador
Barbados	Arab Republic of Egypt
Belgium	El Salvador
Benin	Ethiopia
Bolivia	Fiji
Botswana	Finland
Brazil	France
Bulgaria	Gabon
Burma	German Democratic Republic
Burundi	Federal Republic of Germany
Byelorussian SSR	Ghana
United Republic of Cameroon	Greece
Canada	Grenada
Cape Verde	Guatemala
Central African Empire	Guinea
Chad	Guinea-Bissau
Chile	Guyana
China	Haiti
Colombia	Honduras
Comoros	Hungary
Congo	Iceland
Costa Rica	India

Indonesia
Iran
Iraq
Ireland
Israel
Italy
Ivory Coast
Jamaica
Japan
Jordan
Kenya
Kuwait
Lao Republic
Lebanon
Liberia
Libyan Arab Republic
Luxembourg
Madagascar
Malawi
Malaysia
Mali
Malta
Mauritania
Mauritius
Mexico
Mongolia
Morocco
Mozambique
Namibia
Nepal
Netherlands
New Zealand
Nicaragua
Niger
Nigeria
Norway
Pakistan
Panama
Papua New Guinea
Paraguay

Peru
Philippines
Poland
Portugal
Qatar
Romania
Rwanda
Saudi Arabia
Senegal
Seychelles
Sierra Leone
Singapore
Somalia
Spain
Sri Lanka
Sudan
Surinam
Swaziland
Sweden
Switzerland
Syrian Arab Republic
Tanzania
Thailand
Togo
Trinidad and Tobago
Tunisia
Turkey
Uganda
Ukrainian SSR
United Arab Emirates
United Kingdom
Upper Volta
Uruguay
USSR
Venezuela
Viet-Nam
Yemen
Yugoslavia
Zaire
Zambia

UNITED NATIONS ENVIRONMENT PROGRAM [UNEP] INTERNATIONAL REGISTER OF POTENTIALLY TOXIC CHEMICALS [IRPTC]

The United Nations Environment Program (UNEP) was established by the United Nations General Assembly in 1972 on the recommendation of the Conference on the Human Environment [Resolution no. 2997 (xxvii) 1972]. Acting on the above recommendation, the Governing Council of UNEP, at its second session in 1974, authorized the executive director to convene an expert workshop on the International Register of Potentially Toxic Chemicals. This workshop, held in Bilthoven, Netherlands, in January 1975, established guidelines for the setting up and operation of the IRPTC.

The Governing Council, at its third session in 1975, authorized the executive director of UNEP to establish a Program Activity Center for the International Register of Potentially Toxic Chemicals, to serve as an essential tool in optimizing the use of chemicals for human well-being and at the same time to provide a global early warning system of undesirable environmental side effects [paragraph 8 of Decision 29(III) of the Governing Council]. A Task Team of 13 experts met in Nairobi in July/August 1975 and developed the strategy set out in the Bilthoven report into modes of operation, organizational procedures, and a plan of action.

At its fourth session in 1976, the Governing Council decided that the IRPTC should be a component of Earthwatch, the Global Environment Assessment program of UNEP. Earthwatch consists of four components, namely, Evaluation and Review, Research, Monitoring, and Information Exchange. IRPTC, along with the International Referral System (IRS) for sources of environmental information, forms the Information Exchange component of Earthwatch.

The Governing Council, at its fifth session in 1977, approved 21 goals to be achieved by UNEP by 1982; one of these goals is that IRPTC should by then have the capability to issue warnings and technical publications (on chemicals). It also included IRPTC among the eight topics on which in-depth studies should be presented at its sixth session.

This study recommended, among other things, a revised set of objectives and strategies for IRPTC; the Governing Council, and its sixth session in 1978, approved these objectives and strategies. These are:

To facilitate access to existing data on the effects of chemicals on man and his environment, and thereby contribute to a more efficient use of national and international resources available for the evaluation of effects of chemicals and their control;

On the basis of information in the register, to identify the important gaps in existing knowledge on the effects of chemicals, and call attention to the need for research to fill those gaps;

To identify, or help identify, potential hazards from chemicals, and to improve the awareness of such hazards;

To provide information about national, regional, and global policies, regulatory measures and standards and recommendations for the control of potentially toxic chemicals.

The IRPTC Program Activity Center started its operations in February 1976 from UNEP Headquarters in Nairobi; the offices were transferred to the WHO. The staff of IRPTC at present consists of the director, the chief of the Scientific Program Unit, the chief of the Information Processing Unit, an administrative officer, three secretaries, and an assistant librarian; two scientists are under recruitment.

Within UNEP, IRPTC works closely with IRS and GEMS (Global Environmental Monitoring System)—both of which are also components of Earthwatch, the Environmental Law Unit, the Industry and Environment Office, the Task Force on Pollution and Human Health, and the Regional Seas Program Activity Center.

The major objective of the IRPTC is to collect and publish data on the hazardous effects of chemical substances. The major research projects include the following: to identify the characteristics of potentially hazardous chemicals and to define hazardous characteristics and programs in the data bank. The major publications of the IRPTC include IRPTC attributes for a chemical data register and data profiles of chemicals for the evaluation of their hazards to the environment of the Mediterranean Sea.

WHO Building, Room L31
1211 Geneva 27, Switzerland
Telephone 91-21-11
Dr. J. W. Huismans, Director

APPENDIX K: NATIONAL INSTITUTE FOR OCCUPATIONAL SAFETY AND HEALTH

The National Institute for Occupational Safety and Health (NIOSH) is the principal federal agency engaged in research to eliminate on-the-job hazards to the health and safety of America's working men and women. Part of the U.S. Department of Health, Education and Welfare, NIOSH is headquartered in Rockville, Maryland. Its research facilities are located in Cincinnati, Ohio, and Morgantown, West Virginia. The Institute has five major responsibilities:

Conduct research on the impacts of industrial chemicals, physical factors within the industrial environment, and industrial processes on human health and safety.

Perform surveillance of the effects on human health and safety of actual conditions on the job.

Disseminate to both employees and employers information on the hazards to safety and health from occupational exposures to various chemical and physical influences, on how possible problems can be detected, and on the most effective means for avoiding or controlling the effects of these hazards.

In conjunction with the Occupational Safety and Health Administration of the U.S. Department of Labor, provide training so that necessary personnel are available to safeguard employee health and safety.

Prepare and transmit to OSHA recommendations on criteria for recognition of adverse effects by chemical and physical agents on the health of employees and on the limits that should be promulgated as mandatory standards to control them.

NIOSH publishes and distributes a wide variety of documents on various occupational safety and health topics. These include criteria documents that review the health effects of specific toxic substances, booklets on hazard recognition and and industrial hygiene practices, and an excellent book, <u>Occupational Diseases: A Guide to Their Recognition</u> (DHEW NIOSH

Publication No. 77-181). These materials may be obtained through NIOSH Regional Offices in the following cities:

NIOSH LOCATIONS

Headquarters
National Institute for Occupational Safety and Health
5600 Fishers Lane, Park Bldg.
Rockville, Maryland 20857

Cincinnati Laboratory
National Institute for Occupational Safety and Health
Robert A. Taft Laboratories
4676 Columbia Parkway
Cincinnati, Ohio 45226

Appalachian Laboratory
Appalachian Laboratory for Occupational Safety and Health
National Institute for Occupational Safety and Health
944 Chestnut Ridge Road
Morgantown, West Virginia 26505

Region 1: Boston
(Connecticut, Maine, Massachusetts, New Hampshire, Rhode Island, and Vermont)
Regional Consultant, NIOSH
DHEW Region 1
Government Center (JFK Fed. Bldg.)
Boston, Massachusetts 02203
(617) 223-6668

Region 2: New York City
(New Jersey, New York, Puerto Rico, and Virgin Islands)
Regional Consultant, NIOSH
DHEW Region 2, Federal Bldg.
26 Federal Plaza
New York, N.Y. 10007

Region 3: Philadelphia
(Delaware, District of Columbia, Maryland, Virginia, West Virginia, and Pennsylvania)
Regional Consultant, NIOSH
DHEW Region 3
P.O. Box 13716
Philadelphia, Pa. 19101
(215) 596-6716

Region 4: Atlanta
(Alabama, Florida, Georgia, Kentucky, Mississippi, North Carolina, South Carolina, and Tennessee)
Regional Consultant NIOSH
DHEW Region 4
101 Marietta Tower
Atlanta, Georgia 30323
(404) 881-4474

Region 5: Chicago
(Illinois, Indiana, Michigan, Minnesota, Ohio, and Wisconsin)
Regional Consultant, NIOSH
DHEW Region 5
300 South Wacher Drive
Chicago, Illinois 60606
(312) 886-3881

Region 6: Dallas
(Arkansas, Louisiana, New Mexico, Oklahoma, and Texas)
Regional Consultant, NIOSH
DHEW Region 6
1200 Main Tower Building
Room 1700-A

Dallas, Texas 75202
(214) 665-3081

Region 7: Kansas City
(Iowa, Kansas, Missouri, and Nebraska)
Regional Consultant, NIOSH
DHEW Region 7
601 East 12th Street
Kansas City, Missouri 64106
(816) 374-5332

Region 8: Denver
(Colorado, Montana, North Dakota, South Dakota, Utah, and Wyoming)
Regional Consultant, NIOSH
DHEW Region 8
11037 Federal Building
Denver, Colorado 80294
(303) 837-3979

Region 9: San Francisco
(Arizona, California, Hawaii, and Nevada)
Regional Consultant, NIOSH
DHEW Region 9
50 United Nations Plaza
Room 231
San Francisco, California 94102
(415) 556-3781

Region 10: Seattle
(Alaska, Idaho, Oregon, and Washington)
Regional Consultant, NIOSH
DHEW Region 10
1321 Second Avenue (Arcade Bldg.)
Seattle, Washington 98101
(206) 442-0530

NAME INDEX

Abrams, H., 76
Ahlberg, Rolf, 41
Aimes, C., 51
Al-Aidroos, Karen, 69
Alvarez, Robert, 103
Amor, Adlai, J., 179
Ashford, Nicholas A., xii, 118
Atherley, Gordon, 65
Austin, David, 150
Avery, Robert, 113

Baez, Armando P., 75
Baker, Robin, 83
Baldock, David, 51
Barhad, Bernard, 34
Baron, Sherry, 152
Bassow, Whitman, 139
Beaudry, René, 72
Bedrikow, Bernardo, 168
Bellow, Bonnie, 147
Benedetti, Jean-Louis, 71
Beritic, Tihomil, 57-59
Berlin, Alexandre, 31
Bernstein, Ellen, 78
Biocca, Marco, 28
Blasco, Delmar, 4
Bordman, Joan, 80
Boudreau, Emile, 72
Bruch, J., 25
Brunt, Melanie, 77
Buchberger, J., 39
Bull, David, 54, 121
Burke, Tom, 51

Carter, Arthur, 77
Castorina, Joe, 151
Cezar, Nancy E., 169
Chew, P. K., 13
Chowdhury, Zafrullah, 6

Christenson, John, 94
Cohen, Jamie, 160
Cohn, Gregory, 54
Coling, George, 107
Colohan, A. B., 63
Conwell, Donald J., 156
Coye, Molly, 78
Craig, Marianne, 48, 57
Crawford, W. A., 16

Dalton, Alan, 47
Davies, Clarence J., 27
Davis, Devra, 100
Davis, Morris, 77
Deang, Ricardo T., 175
Deuber, A., 39
Dunlop, Louise, 101, 102

Egleson, Nick, 147
Elkiss, Helen, 110
Elling, Ray, 87
Engler, Rick, 156

Feder, Gene, 48
Fenton, Thomas P., 7, 137
Fluss, S. S., 40
Frankel, Maurice, 55
Fren, James, 173

Garrison, Karen, 78
Gee, David, 46
Gish, Oscar, 124
Glasser, Melvin, 124
Goethe, H., 26
Goldsmith, Frank, 132, 144
Gomez, Manuel, 133
Gordon, Larry J., 95
Greenberg, Michael R., 127
Greenstreet, Susan, 173

Greuter, W. F., 39
Grieco, Antonia, 28
Grgic, Z., 58
Grossman, Rachael, 84
Grossman, Richard, 100
Grover, Margaret L., 128
Guha, Pradeep, 10
Guillemin, M., 38
Guion, Lee, 150
Guise, Jane, 120
Gurzinski, Ellen, 140
Gusman, Sam, 27
Guttmacher, Sally, 130

Haddad, Ricardo, 168
Hagglund, George, 163
Harmon, Suzanne, 97
Harris, LaDonna, 128
Hartgerink, M. J., 32
Henao, Samuel, 167
Hernberg, Sven, 23
Highland, Joseph, 99
Hine, Carole, 145
Hrikko, Andrea, 77
Huismans, J. W., 39
Hunt, Vilma R., 153
Hunter, W. J., 31
Husbumrer, Chinosoth, 14

Ibana, Antonio Granda, 18
Irwin, Frances, 27
Ison, Terence George, 64
Izmerov, N. F., 179

Jasanoff, Sheila, 132
Jerabek, Sandra, 106
Jolley, Linda, 67

Kamlet, Kim, 106
Kaloyanova, F., 20
Kasperson, Roger E., 115
Key, M. M., 161
Khogali, Mustafa, 61
King, Edward, 53
Knight, Jeffrey, 79

Kundig, S., 39
Kupchik, George J., 133

Laden, Vicki, 163
Lambert, Colin, 67
Lappé, Frances Moore, 80
Lascoux, A., 25
Lauwerys, R., 19
Lesage, Michael, 74
Lessin, Nancy, 120
Lin, Vivian, 78, 82
Locke, John, 43
Love, Marsha, 136
Lund, John, 152

Magelli, Leopoldo, 29
Mancs, John, 45
Mancuso, Thomas F., 154
Manu, Petru, 35
Marri, Gastone, 30
Mastromatteo, E., 66
Mazorra, Maria, 94
McCann, Michael, 140
McCarthy, T. F., 66
McGowan, Alan, 147
McKechnie, Sheila, 45
McKowne, Michael, 128
Medawar, Charles, 55
Melaine, Pat, 151
Mergler, Donna, 70
Molina, Gustavo, 167-168
Montague, Joel, 115
Moroles, Adrian, 162
Morse, Linda, 78
Mur, Suma P. K., 171
Murphy, D. W., 112
Murtomaa, Markku, 24

Nagin, Deborah Ann, 146
Nantel, Albert, 71
Ng, T., 15
Niculescu, Toma, 35
Norinsky, Marilynn, 142
Norseth, Jor, 32
Norsigian, Judy, 118
Noweir, Madbulitt, 59

Osman, Yousi F., 5
Osorio, Jorge R. Fernandez, 74
Oyanguren, Hernam, 166
Oyeka, Ike C. A., 125

Painter, Kathy, 106
Pallemaerts, Marc, 21
Panicker, P.V.R.C., 10
Pekka, Nvorteva, 22
Pelletier, Thomas H., 126
Perley, Michael, 68
Peters, Ronald J., 111
Petersen, Borbe, 33
Polakoff, Phillip, 86
Pope, Carl, 84, 148
Prossin, Albert, 64
Pupo-Nogueira, Diego, 164

Raman, V., 10
Ramos, Jovelino, 149
Rath, Amitav, 11
Riegert, A. L., 62
Ringen, Knut, 105
Russel, Michael, 158-159
Rustum, Roy, 153

Sakabe, Hiroyuki, 11
Sampson, Syl, 158-159
Samuels, Sheldon W., 93
Sandalls, Helen, 99
Savoie, Jean-Yves, 71
Sayer, John, 8, 137
Schwartz, Adolph E., 154
Selikoff, Irving J., 134
Seoule, G., 51
Sessions, Jim, 158-159
Shapiro, Helen, 134
Shinoff, Mary, 82
Sidbury, Anne, 106

Sidel, Victor W., 133
Silver, George A., 87
Simches, Sherri A., 92
Slanicka, C. J., 151
Soderstrom, Ingvvar, 37
Soemarwoto, Ir. Otto, 170
Spomenka, Telisman, 58-59
Stellman, Jeanne M., 131
Stewart, Anna, 17
Swomley, James A., 137

Tait, Nancy, 56
Tedford, Bette, 145
Thompson, L. J., 174
Tinker, Irene, 104
Trachtman, Lester, 92
Trudel, Serge, 73

Valesquez, Joe, 115
Van Loon, Eric, 122
Vargas, Miguel Arenas, 75
Vera Vera, Manuel, 157
Verbrugge, G., 19
Vertucci, Christine, 8, 137
Viklund, Birger, 36
Villavicencio, Veronica, 177
Visnja, Karacic, 59

Yuk-King, Cheung, 9

Walters, Elli, 103
Ward, Barbara, 52
Wasserman, Marcus, 60
Watermann, Friedrich, 27
Wegman, David, 116
Weiner, Alan, 150
Whitaker, C. J., 42
Whitehead, Cynthia, 26
Witt, Matt, 95
Wolfe, Sidney M., 107
Wright, Ian, 48

SUBJECT INDEX

accidents, industrial, 37; prevention, 27, 42, 111, 165, 166
acrylonitrile, 148
advertising, 55
affirmative action, 144, 158
Africa, Central, health care delivery, 115
African, mine workers, 93; trade unions, 93
Agency for International Development, health care project, 115
agrarian reform, Chile, 81
agribusiness, 51
agricultural chemicals, 175; commodities, 176; health, 16, 79; industries, 124; practices, 176
agriculture, 114
agro pesticide dealers, 176, 177
air, 89, 101; conditioning, 42; pollution, 146; pollution, Philippines, 178; pollution, Sudan, 5; pollution, USSR, 180; quality, 137
alcoholism, 64
allyl chloride poisoning, 7
aluminum industry, 155
aluminum reduction workers, 62
American Indians, 129, 130; coal, 129; minerals, 129; U.S. government, 129
Americas, 134
appropriate technology, 125, 126

aquatic plants, 22
Army Corps of Engineers, 101
arsenic, 117, 148
asbestos, 47, 72, 136, 148, 154, 157, 162, 167, 169; cement materials, 157; dust, 63; exposure, 29, 86, 117, 172; hazards, 56, 150; studies, 16; usage, 25; toxicology, 167; union code of practice, 172
asbestosis, 167
Asia, 138
asthma, 34, 62
atomic energy, 15, 154
audiometric technicians, 62
audio-visual production, 148
Austria, 95
automobile, 124
automotive workers, 165
auto workers, 94

back care, 163
Bangladesh, 81
banking policies, 51
benzene, 148, 165
benzidine, 154
beryllium, 154
betanaphthylamine, 154
Brazil, 94
bronchitis, chronic, 167
building materials, 55
Bureau of Reclamation, 101
byssinosis, 43, 61

cadmium, 53, 165, 172
Canada, 69
cancer, 42, 46, 56, 68, 70, 109, 113, 127, 154, 172

302

carbon black exposure, 117
carbon monoxide, 117, 169
cardiovascular investigations, 16
caustic soda, 22
cement, 60, 157
chemicals, 27, 29, 40, 77, 114, 117, 164, 171, 175, 177, 178
child health, 88
Chile: agrarian reform, 81; industry, 169
chlorine exposure, 117
chromate, 154
chrome, 165
chromium, 62
Citarum River Basin, 171
citizen participation, 147
civil rights, 58
clerical workers, 79
clipping services, 82
coal, 102; American Indians, 129; dust exposure, 117; dust studies, 64; energy source, 102; leasing, 102; miners, 117, 167; mining services, 64
coastal resources management, 98
coastal zone: management, 178; preservation, 146
collective bargaining, 113, 124, 158
Colombia, 94
community health programs, 173
community organization, 110
compensation, 68, 108
computers, 147
conservation, 79, 101, 108
construction industry, 77, 169
construction safety, 62; worker, 56
consumer policy, 118; affairs, 113

continental shelf oil and gas, 102
contract language, 87, 155, 156
copper smelting, 144
coronary heart disease, 15
corporate research, 67; responsibility, 141
corporations, U.S., 143
cotton dust, 60
crops, 176
culinary workers, 82
cytogenetic monitoring, 161

deep seabed mining, 102
Depoprovera, 6
dermatitis, 164; contact, 19; epoxy resins, 16
developing countries, 114, 116, 125, 135; occupational health, 61; occupational health problems, 5; technical assistance, 135
development, 52, 54, 121, 125, 149, 169; American Indians, 129; ecology of, 170; women, 104
dimethoate, spray, exposure to, 5
disease, etiology, 130
DNA, recombinant, 79
dust, 25, 29, 46, 56, 70, 180; control, 16; exposure, 60; studies, 16
drugs, 7, 24

Earthscan, 53, 106
economic development, 129, 141
education, 155
elderly, 27
electronics industry, 83, 84
employment, 97, 155
energy, 52, 79, 96, 101, 108, 145, 146; alternative technologies, 101; conserva-

[energy]
tion, 98, 101, 102, 103; development, 101; economic growth, 141; facility siting, 102; forecasting, 102; health effects, 162; policy, 109; programs, 49, 104, 105; research, 20; soft energy, 79; studies, 11; technologies, 100; workplace injury, 144
engineering factories, 34
engineering and system safety, 85
environment, 127, 129, 132
environmental carcinogenesis, 114
environmental exposures, 29; ethics, 98; health, 118, 133, 137, 153; impact assessment, 170, 178; issues, 153; laws, 89, 99, 109; organizations, 19; planning, 116, 127; policy, 89; problems, 52, 169; protection of, 68, 69, 129, 178; quality, 68, 178; regulation, 132; toxicology, 60
epidemiology, 33, 34, 154, 172, 177; occupational, 161; post-graduate, 32; studies, 29, 67
ergonomics, 16, 23, 26, 32, 37, 164
ergophthalmological problems, 26
etiology, 180
executives, 117
export-import bank, 101

factory: inspectorates, 135; workers, 69, 74, 172
fair housing, 109
family planning, 156
farm workers, 94, 162
federal coal leasing, 102
Federal Power Administration, 101
fertilizer, 22, 175
firefighter, 117
fire protection law (Swedish), 37
first aid training, 62, 63
fish, 22, 176
fissous zeolytes, 162
flax dust, exposure, 60
food, 114; policy group, 81; processing industry, 162; regulation, 108; self-reliance, 122
forest land, 129; resources, 171; work, 27, 39
fossil fuel, 101
foundries, 77

garment workers, 82
gas, 39, 70, 102
G.A.T.T., 36
genetic engineering, 122
granite workers, 117
grievance procedure, 113

hazard, awareness, 136, 140
hazard, electrical, 75; identification, 117; occupational, 124; recognition, 153; recognition training program, 151; surveys, 14; technological, 116
hazardous, environments, 36; products, legislation (Swedish), 37; technologies, 20; waste, 106, 107, 145; waste management, 107
Health and Safety at Work, 45
health care: children, 32; delivery systems, 108; delivery systems (Ivory Coast), 115; developing

[health care]
countries, 109; home, 156; preventive, 32; primary training materials, 115; third world, 94
health education, 119
health hazard: chemical, 37; evaluations, 124; recognition, 162
health maintenance organization, 130
health network, 42
health policy, 33
health promotion, 96
hearing protection equipment, 67
heat, 39, 157, 167; research, 61; stress, 19, 136
hemp dust, 117
herbicides, effects of studies, 16
housing, 155, 157
human environment management, 171
human rights, 20, 121
hunger, 80
hypertension, 15

immigration, 149
India, 104
Indian resource development, 129
industrial: accidents, 37; chemical exposure, 19; dermatitis, 62; development, health effects, 179; epidemiology, 11; hygiene, 13, 20, 23, 45, 164, 165, 171; hygienists, 150; medicine, 33; microclimate, 180; physiology, research, 11; pollutants, 75; productivity, 118; relations, 135; safety, 23, 84, 131, 148; siting planning, 27

industries: hazard, 155, 165; small, 165; training, 77
International Conference on Exportation of Hazardous Industries to Developing Countries, viii
irrigation, 129
Israel, 60

labor: history, 113; law, 113; litigation, 72; outreach work, 78; research institute, 152; studies, 112; unions, 149
ladder safety, 163
land, 89
land use, 52, 101
lead, 19, 140, 162, 167, 172; absorption, 53; contamination, 167; exposure, 57, 58, 117, 169; levels, 165; poisoning, 16, 43, 58, 62; studies, 16; toxicology, 167
leather workers, 117
legal services, 78, 156
legislation, 35; European, 31; human rights, 121; international, 40; national, 40; safety and health, 42; Swedish, 37
library information service, 166
lichens, 22
lifting, 16
liquified petroleum gas, 102
livestock, 176
loading, 62
lung disease, 136; adult, 137; coal miners, 168; occupational, 137, 168; pediatric, 137

machinery guarding, 42
manganese, 165, 169
marble exposure, 117

materials handling, 16
meat packers, 117
medical: programs, 74; services, 156, 173
mental health, 143
mercury, 19, 22, 76, 167, 169
metals, 16, 38, 69, 70, 171
methane, 147
microcystic bloom, 170
micro-electronic technologies, 51
migrant workers, 5
mineral, 129
mineral oil, hazards, 16; resources, 102
miners, 156
mining, 11, 102, 129, 162
Mozambique, 81
multinational corporations, 11, 17, 138

national health insurance, 143
National Health Service, England, 44
National Labor Relations Act, U.S., 132
native Americans, 155
nature conservation, 170
natural resources, 27
nautical medicine, 26
New Zealand, unions, 172
noise, 19, 26, 27, 37, 39, 42, 46, 47, 70, 87, 89, 157, 162, 164, 167; abatement, 27; audiovisual, 68; disease, 180; exposure, 117; legislation, 70; metallurgical workers, 165; textile industry, 60
nuclear: energy, 101, 102, 103; power, 122, 150; power plants, 127, 142; power plant safety, 123; siting, 101, 102, 103; waste, 103
nutrition, 96, 122

ocamylase, 43
occupational: accidents, 23, 42; allergens, 43; disease, 11, 13, 14, 18, 23; exposure, 14
occupational health, 87, 133, 137, 142; law, 72; news, 82; nurses, 166, 173; problems, workers, 70, 140; services, 30; standards, 153
occupational safety engineering, 166
occupational safety and health, 141, 148, 151, 177; Act of 1970, 90, 123, 164; Act of Puerto Rico, 157; Act of Sweden, 36; Asia, 8; bilingual program, 162; collective bargaining, 132; committees, unions, 169; cultural relevance, 80; economic problems, 70; enforcement, 141; information, 65, 144; laws, 160; legal aspects, 68; policy, 142; problems, 77; statistics, 65; supervision, 166; symposium on standards, 66; training, 35, 111, 166; worker initiatives, 30
oil, 34, 102, 147, 157, 162
organic solvents, 29
organochlorine compounds, 60
ozone, 117, 120

paints, 41
paper industry (Sweden), 37
pediatrics, 179
personal protective equipment, 42, 131
pesticides, 5, 16, 46, 79, 89, 94, 160, 161, 162, 163, 165, 168, 171, 175, 176, 177, 178
Philippines, 104, 175, 176, 177, 178

physicians, 172
pipecovers, 117
plastics, 41
pneumoconiosis, 25, 64, 165, 167
poison, 71
political economy, 134
politics and health, 130
pollutants: workers, metal refinery, 69
pollution: biological ecosystems, 75; compensation, victims, 68, 133; environmental, 55
population, 116
Portugal, 94
powerline siting, 102
primary health care, 143
printers, 82
product safety, 88, 108
public health, 20, 32, 44, 176
Public Health Administration, 32
public health training, 116, 117
public lands, 79
public media center, 82
pulmonary disease, occupational, 61
pulp industry (Sweden), 37
pumping equipment, 37

radiation, 16, 37, 40, 46, 69, 89, 101, 102, 103, 154, 157, 163
radioactive waste management, 116
rayon industries, 154
recreation, 129
regulation, 147
renewable energy sources, 114
reproduction, studies, 69
reproductive damage, 70; hazards, 157; rights, 97, 146

rescue procedures, 42
respiratory: diseases, 16, 34, 43, 137, 156, 180; protective equipment, 67
risk assessment, 132; prevention, 42, 111
road safety, 37
rubber, 151, 154
rural land, 103

SALT, 101
screening clinics: (for) occupational diseases, 150; textile workers, 150
seamen, 173
sewage, 76
shift work, 37, 46
shipbuilding, 26
shipyard workers, 163, 169
shoe industry, benzene exposure, 165
silica, 29, 63
silicosis, 7, 167
skin disease, 62
smelter workers, 117
smoke, 117
social justice, 158
social medicine, 32
social responsibilities, 55
social service programs, 75
sociobiology, 122
soil erosion control, 178
solar power, 101
solid waste, 89, 108
solvents, 16, 38, 39, 41, 70, 167, 169
southern exposure, 149
Spain, 94
stress, 87, 117
strip-mining, 102
styrene, 39
sulfur dioxide exposure, 117

Tanzania, 81
TDI, exposure, 117
technological innovation, 118

technology, 11; policy, 118; women, 104, 105
technology transfer, 125
Tennessee Valley Authority, 147
tetra chlorodibenzodioxins, 29
textile industry, 148
textile mills, 5
Third World, 113, 124, 149
timber management, 141
tourism, 129
toxic: chemicals, 27, 55, 97, 99; metals, 38
toxicology, 16, 20, 23, 33, 71, 167, 180
toxics, 106, 135
toxic substances, 20, 27, 42, 77, 89, 101, 108, 109, 127, 132, 133, 146, 178
toxins, international, 146; national regulations, 148
trade secrecy, 108
trade unions, 45, 47, 54
training: safety representatives, 47; stewards, 113; union officials, 163
transportation, 34, 52, 109, 117, 146
trichloroethylene, 157
tunnel workers, 117

UNITAR, 138
United Nations Center on Transnational Corporations, 138
United Nations Conference on Science and Technology for Development, 104
United Nations Economic and Social Commission for Asia and the Pacific, 177
unemployment compensation, 113

ventilation, 42, 46, 87, 157
vibration, 26, 42, 47, 62, 67, 180
vinyl chloride, 117
visual display units, 39, 46

waste, 108
waste alert program, 106
waste, nuclear, 147, 148
waste, transportation of, 101
wastewater treatment, 75
water, 101; conservation, 101, 103; pollution, 10, 75, 146; quality, 76, 98; resources, 99, 145; resources policy, 102; resources management, 171; resources policy, 102; supply, 26, 114; systems, 89; transportation, 101; user conflicts, 101
welders, 117
West Germany, 95
white lung, 150
wildlife, 52, 79, 99
women, 120, 140, 143; development, 104; food technology, 104; health, 104, 119; occupational health, 57, 69, 87, 131; poverty 105; power, 104; pregnancy, 7, 70; productivity, 172; Southeast Asia, 84; technology, 105; work hazards, 153
wood products, 95
woodworkers, 95
worker: compensation, 62, 64, 113, 120, 132, 157, 158; education, 33, 152; environment, 164; health, 67, 69, 82; information and resource center, 70; minorities, 109; notification programs, 86; rights, 147, 158, 162; women, 70

workplace conditions: chemical risks, 29; cadmium smelting, 143; lead smelting, 143; zinc smelting, 143
workplace, problems, 164
workplace, quality, 113
workplace, safety and health, 108
workplace, stress, 164
worksite, inspection, 62
world hunger, 80, 81

ABOUT THE AUTHOR

JANE H. IVES is a native of Newton, Massachusetts. She holds an M.Sc. degree from the London School of Economics, England, and she is a doctoral candidate at the University of London, England. Currently she is an international consultant. This catalogue was initiated at the International Conference on the Exportation of Hazardous Industries, Technologies and Products to Developing Countries held at Hunter College, New York, N.Y., November 1979 which Ms. Ives conceived, organized, and chaired. She has testified before the U.S. Congressional Committee on Foreign Affairs on the exportation of hazardous industries, technologies and products and she has also delivered papers on this subject throughout the United States and Europe.

NO LONGER THE PROPERTY
OF THE
UNIVERSITY OF R. I. LIBRARY